SURVIVAL IN THE WILD

CAMOUFLAGE AND MIMICRY

Denis Owen

CAMOUFLAGE
AND MIMICRY

CAMOUFLAGE AND MIMICRY

Denis Owen

THE UNIVERSITY OF CHICAGO PRESS

The University of Chicago Press, Chicago 60637

This book was designed and produced by
The Rainbird Publishing Group Limited
40 Park Street, London W1Y 4DE

House Editors: Karen Goldie-Morrison Linda Gamlin David Burnie
Design: Rod Josey Ltd
Production: Clare Merryfield

89 88 87 86 85 84 83 82 1 2 3 4 5

Library of Congress Cataloging in Publication Data

Owen, Denis Frank.
  Camouflage and mimicry.
  First published in U.K. by Oxford University Press, 1980.
  (Survival in the wild)
  Bibliography: p. 150.
  Includes index.
  1. Camouflage (Biology) 2. Mimicry (Biology)
I. Title. II. Series: Survival in the wild.
QH546.093 1982   574.5'7   82-2566
ISBN 0-226-64188-0 (pbk.)   AACR2

Photosetting by SX Composing Limited, Rayleigh, Essex
Illustration origination by Hongkong Graphic Arts Service Centre
Printing and binding by South China Printing Co. Hong Kong

Front cover: The extraordinary eye-spot display of the
South American frog *Physalaemus nattereri*. Its backside
is raised towards the attacker who is frightened
away by the appearance of the 'eyes.' *I. Sazima.*

Contents

Oh, what a tangled web we weave,
When first we practise to deceive!

Sir Walter Scott,
Marmion, 1808

Foreword

A full understanding of animal species can be acquired only after years of extensive studies in their natural environments. Only in the wild is it possible to discover the evolutionary and adaptive significance of each biological activity. It then becomes apparent that many of the forms, colours and activities of wild animals and plants are adaptive responses to the basic problems of survival: the need to eat, to avoid being eaten, and to mate and reproduce. Each species is beset with a unique set of problems depending on the type of environment in which it lives and on its structure: whether it is in a desert or in a jungle, or whether it is a frog, a tiger or a fly. Each species has evolved its own repertoire of strategies which enable it to survive. A successful individual not only survives but also reproduces and passes on its genes to the next generation. However, only those individuals best adjusted to their environment survive, and they transmit the traits which have made survival possible to their offspring, an idea embodied in the phrase 'survival of the fittest'.

It is the aim of this new series *Survival in the Wild* to describe and explain the bewildering diversity of strategies displayed by the living world. Each book selects a biological activity vital to survival and describes the array of physical and behavioural adaptations which have evolved as a result of fierce competition. In an often hostile world, individuals interact with others, as food sources, or potential predators to be avoided, or mates.

Camouflage and Mimicry shows how, by using a combination of colour and shape, many animals provide themselves with elaborate and highly effective disguises. Some have evolved uncanny likenesses to elements of their environment such as leaves, rocks and snow to avoid being noticed by predators or prey, while others use colour and shape in spectacular displays to frighten a predator or to warn it that they are poisonous. Some edible animals have taken advantage of warning coloration and have evolved stunning likenesses to other, poisonous animals. Such mimicry is employed by several groups of vertebrates, including birds, snakes, and salamanders, but it is most widely and brilliantly displayed among the insects where it is sometimes only the experienced naturalist who can see through the deception.

Denis Owen spent several years in Africa, particularly in Uganda and Sierra Leone, unravelling the complex mimetic relationships between several families of tropical butterflies. His interest in the subject is clearly revealed in his portrayal of these families and the astonishing degree to which the mimicry has evolved. But this is only part of his text. He begins by setting the scene for the imitation of the natural world with a discussion of the wider aspects of camouflage. In the green and brown world, animals have evolved to resemble leaves, grass, twigs or flowers. In desert regions, both plants and animals may look like stones or rocky ground. He then moves on to a discussion of polymorphism in invertebrates which he suggests is another ploy for avoiding recognition as possible prey. Finally, he tackles warning coloration and the most perfect deceptive strategy of them all, mimicry. The examples he uses become increasingly ingenious. For example, he describes an American passion flower plant, the tips of whose tendrils look like the eggs of a butterfly. The plant thereby avoids the attentions of egg-laying females whose larvae would soon damage it. But, however fantastic the examples may seem, Denis Owen never fails to emphasize the importance of experimental work in elucidating the functions of polymorphisms, of camouflage patterns or of mimetic forms. He explains why much of this work has involved insects and other invertebrates because they are comparatively easy to keep and breed under laboratory conditions. It becomes clear that many vertebrate mimetic relationships must remain conjecture until they can be fully tested, although they are interesting all the same.

In the course of writing this book, Denis Owen has obtained valuable information from many people, either in conversation or in their published work in books and journals. He particularly thanks John Allen, David Blest, W. S. Bristowe, Lincoln Brower, Arthur Cain, Dennis Chanter, Sir Cyril Clarke, Lawrence Cook, Harvey Croze, Dave Duthie, Malcolm Edmunds, Thomas Eisner, G. B. Moment, Miriam Rothschild, Gustaf Rudebeck, L. de Ruiter, Ted Sargent, John Sellick, Allister Smith, David Smith, John Turner, Derek Whiteley, W. Wickler and Christer Wiklund. David Smith also read the manuscript and made some valuable suggestions.

1 The need for deception

No animal is safe. There are predators everywhere — waiting, lurking, running, jumping, burrowing, flying and swimming — in woods, fields, swamps, rivers and lakes; in the depths of the ocean and high on mountains. If an animal is seen, heard or smelt it is potentially in danger. The possibility of discovery and of subsequent death provides the arena for the evolution of deceptive strategies. Animals may be camouflaged by a coloration which matches their surroundings, or, by a combination of colour and shape, they may mimic another animal.

This book is about camouflage and mimicry as both defensive and offensive strategies. It is mostly about animals that have evolved ways of deceiving predators. But predators, too, use deception as a means of gaining access to prey, and there are some plants with deceptive strategies similar to those found in animals. Although the book is called *Camouflage and Mimicry* I shall from time to time digress and discuss other means of defence and offence. Some animals are unpalatable or poisonous. Their protective strategy is not exactly deceit; indeed they often make themselves conspicuous and rely upon bold, bright coloration as a means of warning off would-be predators. This is known as warning coloration and without it many forms of mimicry would not be possible.

The animal in its environment

No animal is safe but it would be wrong to give the impression that animals spend their lives doing nothing but indulging in deception. They must also feed, escape the hazards of bad weather and, above all, reproduce successfully. Deception is just one of a set of strategies that increases an individual's chances of survival and of transmitting its attributes to the next generation. Each species has its own particular set of strategies, which are adapted to the environment in which it lives, including the other animals with which it interacts. The study of how an animal is adapted to its environment forms part of the science of ecology, and certain important concepts in ecology, that of a niche, of a community and of a food web, must be explained before we go on to look at deceptive strategies in particular.

The word niche is used to describe an organism's role or 'profession' in the environment in which it lives. By specifying an individual's niche we not only describe when and where it lives but how it operates in relation to other individuals and species present, and particularly how it obtains food and shelter, escapes enemies and breeds successfully.

A butterfly may spend much of the day seeking flowers from which to obtain nectar. It constantly 'refuels' from the energy-rich, sugary nectar and recognizes, by colour and scent, flowers of the right structure into which it can insert its sucking tongue. Some flowers are unsuitable while some do not produce nectar, and these are avoided because it is a waste of time to try and exploit them. It is also a waste of energy and there is no point in taking unnecessary risks with spiders and mantids which lurk in flowers and seize butterflies by the head.

The butterfly recognizes possible mates by colour and scent and may produce its own scent to attract a mate. A female butterfly is an expert botanist and rarely lays eggs on plants which the resulting caterpillars would be unable to eat. Potential enemies are recognized and appropriate evasive action is taken. During the course of a day a butterfly may experience repeated encounters with competitors, enemies and mates, and because it is genetically programmed to make the right responses it will, unless it makes a mistake, live to see another day.

By operating effectively in its niche the individual survives and reproduces. If displaced from its niche it often fails: a tropical swallowtail butterfly would simply not last in an English woodland and nor would an English peacock butterfly last in a tropical forest. Each has evolved adaptations that enable it to exploit the resources and cope with the hazards of its own environment.

The assemblage of species that occurs in a particular place is called a community. This term should not be confused with what people refer to as a human community in which individuals work together (or supposedly work together) for the common good. A biological community is intensely competitive; there is danger everywhere: from predators and from the ever-present risk of failing to find food and shelter, and failing to find a mate.

A community must be understood in terms of the number of species present and how abundant each of them is, and in terms of the interactions between individuals and species: it is a matter of who eats whom and who breeds with whom. Some communities, such as tropical forests and coral reefs, are exceedingly rich in species, but others contain few because it is too cold or too dry or because some other limiting factor restricts the number of species that can exist. Ponds are rich in species of plants and animals but hot springs contain few because the temperature of the water prevents most from surviving.

As soon as green plants establish themselves a community begins to form and animals of one sort or another follow. Plants compete for space, water, nutrients and light, and their abundance and diversity depends on the rate of supply of these commodities. Each species differs in its requirements, and exactly the same is true of animals, whether they are plant-feeders or predators. Indeed it is an axiom that no two species of organism have exactly the same source of food, optimum conditions for mating and successful rearing of young, or other requirements for survival. In others words the niches of different species are distinct, although in some cases there may be considerable overlap. If we examine a group of similar-looking species of butterflies we find that each differs in the way it exploits nectar sources, in the food-plants of the caterpillars, the time of year at which the adults fly, and the ways in which all stages of the life cycle minimize the risks of being found by predators. This means that a biological community is organized from a basis of struggle and competition: organisms are dependent on resources provided by the environment, but to get what they want they have to compete.

A diagram showing all the possible feeding relationships in a community such as a garden, field or woodland would require an enormous piece of paper. There is a complex food web, ramifying in all directions: everything is eaten by something else, but as a generalization we can construct a

Opposite page A common protective strategy is for an animal to be paler below than above. However, the caterpillar of the European eyed hawk-moth *Smerinthus ocellata* rests upside down and is paler above than below. The top photo shows the caterpillar in its normal resting position in which it appears flattened and leaf-like, while the bottom photo shows what it looks like if perched, unnaturally, on top of the twig.

simple food chain,

plant ——→ plant-feeder ——→ predator

This shows only one predator, but there may be up to four, or rarely five in the chain. The top predator is usually a large animal, for example a bird of prey, lion or shark, which when fully grown is effectively free from predators. But the young of such predators are vulnerable: if left unguarded the nestlings of the most powerful birds of prey are attacked and eaten by crows, and baby sharks are caught and eaten by larger fish.

Life is hazardous. Many animals are killed and eaten by other animals, and many predators die from starvation because they fail to secure prey. Since animals appeared on the earth, predators have continued to evolve adaptations that enable them to locate and kill prey, and prey have evolved adaptations that reduce the risk of being eaten. Everything points to a special kind of arms race with elaborate strategies and counter-strategies for attack and defence. Neither side can be the outright winner. The need for deception on both sides is paramount. Survival is at a premium, and those that do survive reproduce and transmit to their offspring the traits that have made survival possible.

Protective strategies

It is possible to classify the kinds of deceptive and protective strategies found among animals (and among a few plants) likely to be eaten by predators, and at the same time identify the ways in which the predators respond to these strategies. This classification cannot be too precise – predators range from tiny wasps to eagles and lions, while many species fall into more than one category – so a generalized classification is very much a simplification.

1. Camouflage Camouflaged animals resemble living or dead vegetation, soil, rocks and in a few instances the droppings or faeces of other animals. The colours and patterns match the background well either by blending or by resemblance to a specific structure, such as a leaf.

No matter how well an animal is camouflaged there is still the problem of concealing its shape. Animals are three-dimensional and since light normally comes from above the lower part would appear darker, making the animal conspicuous were it not for a common adaptation: nearly all camouflaged animals are paler below than above, a phenomenon known as countershading. Countershading is best seen in fish, mammals and the caterpillars of certain moths. The significance of countershading is that it makes an animal appear shapeless and hence difficult to see. As if to reinforce its importance, animals that habitually rest upside down are paler above than below; these are said to have reverse countershading. There is another solution to the problem of concealing shape: many species are marked with bold stripes or bands which are either much darker or much lighter than the rest of the coloration. Such markings are called disruptive coloration and serve to break up an otherwise distinctive and characteristic outline.

Most camouflaged animals are active at night and remain motionless during daytime: indeed keeping still is probably as important in escaping detection as the camouflage itself. Camouflaged animals are generally palatable and, although often common, they are easily overlooked by the human observer and presumably by their predators. Predators find camouflaged prey by constantly searching and repeatedly capturing those individuals which for some reason do not match their background as well as other individuals; they are quick to detect movements, and learn by experience. Familiarity with the local environment is useful to the predator in finding camouflaged prey. There are a few species of camouflaged plants, and some predators are presumed to be camouflaged to aid them in approaching prey. In many other predators (for example, mantids) the camouflage probably plays

Glaucilla is a carnivorous sea-slug of the Pacific. It is unusual in that it floats with its muscular foot uppermost, and it is this side of the animal which is dark. The surface facing downwards which can be seen by predators swimming below is pale and less easily visible against the light.

a dual role in that it both deceives their predators and conceals them from their prey. Camouflage is described in Chapters 3 and 4.

2. Diverse coloration A few, usually very common species of animals, occur in a variety of contrasting colours and patterns, so that anyone looking at them for the first time would find it hard to believe that they all belonged to the same species. They are normally palatable. Since many predators learn from previous experience, a prey item that stands out or contrasts with others in the population is often missed simply because it looks different and because diversity is confusing. This type of deception may or may not be combined with some form of camouflage. Chapter 5 deals with this phenomenon.

3. Diverting structures and coloration Some animals, mostly insects, have extensions to their bodies which divert the predator's attention from vital to non-

vital parts. The predator strikes but is left with a piece of wing, tail, or similar structure, and the prey escapes with only minor damage. Eye-like markings on the outer edge of a butterfly's wing divert the predator's attention away from the body: a bird which snaps at a butterfly's wing often misses the butterfly. Feigning injury or death is also an effective means of deception in some species, including certain snakes, birds and mammals. These anti-predator devices are described in Chapter 6.

4. Frightening and startling coloration and behaviour Until discovered by a would-be predator, animals with this type of strategy appear camouflaged, but upon being disturbed they expose conspicuous and bright flashes of colour, eye-like markings, or patterns that suggest a resemblance to something unpleasant such as a snake. They may also appear aggressive. The predator is often frightened, startled or intimidated;

To avoid being caught by a predator, a grasshopper must sit still. This grasshopper has been seen by a side-striped chameleon *Chameleo bitaeniatus*.

it hesitates, and while recovering its composure the prey gets away — or is simply left alone. Animals with this type of coloration and behaviour usually remain motionless by day and do not move unless disturbed. They are palatable and frequently avoid being eaten because the predator does not investigate further. A really experienced predator is less likely to be put off by them than an inexperienced one. The various methods used to intimidate predators are described in Chapter 7.

5. Warning coloration There are brightly-coloured and boldly-marked animals which make little or no attempt to hide, keep still, or move away in the presence of danger. Many are white, yellow, red or orange with conspicuous yet simple black markings. They are noxious, even toxic, to predators. They taste or smell foul, sting, bite, irritate, produce unpleasant secretions or mucus, and may cause sickness if eaten. Predators tend to avoid them. The warning coloration advertises their unpleasantness and

The chameleon strikes the grasshopper with its long tongue and, provided it can dislodge it, has made a successful kill.

predators become conditioned to leaving them alone. Chapter 8 deals with warning coloration.

6. Mimetic coloration Some palatable animals resemble warningly coloured species; they have the same or similar bright colours and bold markings and predators that have become conditioned to avoiding noxious animals tend to leave them alone. Mimics are usually relatively rare; if they were common predators would soon learn by experience that they are not really unpalat-

able. There is in addition a special kind of mimicry in which several unpalatable species resemble one another. Mimicry is discussed in Chapters 9 and 10.

A predator's view of prey

First, something should be said about the use of the words 'predator' and 'prey'. A predator can be a predator more than once, but an individual that becomes a prey item necessarily finds itself in this position once only, which leaves us with a semantic difficulty. But rather than repeatedly using the clumsy expression 'potential prey' we shall stick to the word prey on the understanding that we are thinking of an animal that stands a chance of being captured and eaten.

A predator first has to find its prey. In the above classification of protective strategies, categories 1 and 2 reduce the chances of discovery while categories 3–6 reduce the likelihood of being captured and eaten once the prey has been seen. How do predators react to the first two strategies, that is, camouflage and diverse coloration?

Birds are a familiar group of predators, and ones whose behaviour is easily observed. Many feed wholly or partly on insects and other small invertebrates. Most birds, especially the insect-feeders, could be described as 'conservative opportunists'. They learn by experience and consistently seek more of a kind with which they have already had success. For many birds, if a food item has not been tried before either it does not exist, it is frightening, or it should be ignored. Our own attitude to untried food is somewhat similar.

We do not of course know exactly what a bird sees when it is hunting for food. To have this information we would have to have the eyes and the brain of a bird; but there is every reason to suppose that a bird's view of the environment is similar to our own. Like humans, but unlike most mammals, birds have a poorly developed sense of smell and excellent colour vision. For us they are in many ways easier to understand than mammals even though we ourselves are mammals.

If you watch birds seeking food in the garden it is easy to form the impression that they are constantly on the move and that they are after something quite specific. It is rather like our own behaviour when blackberry-picking: there may be blackberries close at hand but there always seem to be better ones a bit further on. Experiments show that birds do seek out items which they know

are edible, and ignore other food items until all such familiar ones are exhausted. Birds form a 'search-image' of the particular food item. It is easy enough to understand for we ourselves form search-images. Indeed once you have been blackberry-picking for a while you become oblivious to anything except blackberries and they become much more noticeable. After a couple of hours you can see blackberries even when you close your eyes.

Predatory mammals lack colour vision and detect prey chiefly by smell and movement. Snakes and lizards have colour vision and although it may not be identical to that of birds they seem to seek and locate prey in much the same way, except that they are much more inclined to wait for the prey to approach them: unlike birds they are not 'blackberry-pickers'. As for amphibians, we know from experiments that toads can learn by experience and become conditioned to rejecting unpalatable items, like bees, that they have tried before. Insects as a group can see further into the violet end of the spectrum than birds and reptiles. What is known about insect colour vision is mostly derived from experimental work on non-predatory species such as honeybees and butterflies and their ability to discriminate flower colours and patterns. We know that mantids and hunting wasps are visual predators, as are certain families of spiders. Wolf spiders and jumping spiders are active hunters which appear to seek their prey. But exactly what sort of view these predators have of prey we cannot really say, except that, as with virtually all predators, movement on the part of the prey is the great give-away.

Deception as a way of life

This book is about camouflage and mimicry, and other deceptive strategies, particularly as they affect predator–prey relationships. It is dominated by examples in which birds are the predators and insects (especially butterflies) the prey. I make no excuse for this. Much of the

Deception at its best. At the top is the North American monarch *Danaus plexippus*, a highly distasteful butterfly. Predators tend to avoid it.
The viceroy *Limenitis archippus* (below), a totally unrelated species, mimics the monarch and so gains protection from predators. (More on this in Chapter 9.)

tegy but they have not been scientifically investigated and so there is no direct evidence one way or the other. Indeed the literature on protective strategies is full of anecdotal examples. I have used some in this book, but on the whole I have concentrated on examples where there is good evidence.

Why, for example, are polar bears white? An obvious answer is that they match the white Arctic landscape in which they live. But polar bears are big carnivores with no natural enemies, apart from man, and so white camouflage could only serve as means of approaching prey more effectively. This could be the answer to the question, but there are other possibilities. White fur might also reduce heat loss from a warm-blooded animal living in a cold climate. In mammals and birds dark fur or feathers are usually due to the presence of melanin pigment. The production of melanin is a complex chemical process and, like all such processes, expends energy. There is no point in using energy unless there is an advantage in being dark, and this alone might explain why polar bears are white. Many burrowing and cave-dwelling animals are paler than those that live in the open and this is presumably no more than an energy-saving adaptation.

And so we have to be a little cautious, especially when it comes to attempting an explanation of a particular kind of coloration which has not been properly studied in the field or in the laboratory. There may be several different functions for this or that colour pattern which may or may not be involved in predator–prey relationships.

Deception as a way of life occurs throughout the animal kingdom and in some plants. Like all adaptations to the environment it results from and is maintained by natural selection, a well-understood process which we take up in the next chapter.

field and experimental work on the subject has been done on these two groups of animals and it is only by doing experiments that we find that we are really dealing with camouflage, warning coloration, mimicry, or some other aspect of a protective strategy. Birds and insects are simply easier to work with than, for example, mammals, lizards, snakes and many of the groups of aquatic invertebrates. There are many, many examples of what appears to be a protective stra-

Natural selection and adaptation

The ability of an individual, whether predator or prey, to survive and reproduce depends on how well it is adapted to the environment in which it lives. All adaptations are genetically controlled and their presence and persistence is the result of the process of natural selection, the most unifying of all biological principles, first put forward by Charles Darwin and Alfred Russel Wallace in 1858. The theory of natural selection depends upon several propositions which can be observed and tested.

The first is that all organisms have an enormous capacity for growth in numbers. This capacity is rarely realized because death rates are high and most populations are regulated by the availability of resources in the community in which they live. If an individual female butterfly lays a hundred eggs these will, on average, produce only one male and one female in the next generation; the losses of eggs, caterpillars and pupae total 98 per cent. Some of the losses result from the actions of predators, others from food shortage, parasites, disease and weather: much depends on the species in question and the environment in which it lives. High death rates are characteristic of all populations: if they did not occur numbers would rise steeply. This occasionally happens when a species invades a suitable new area, but the population is soon regulated by the availability of resources.

The second proposition is that organisms are individually variable — only relatively rarely are two identical — and that variations are inherited. If a random sample of, say, a species of snail is collected from a single locality it is easy to see that individuals differ from one another slightly in size, shape, colour and pattern. The same is true of virtually all species of animals and plants.

The third proposition is that survivors are likely to differ in certain respects from those that die. The survivors will tend to be those which are better at getting away from predators, finding food and shelter, and so on. This is another way of saying that the probability of death among individuals depends on their genetic make-up. Those with characteristics which are well adapted to the environment tend to survive and pass those characteristics on to their offspring. Gradually such characteristics become more and more common in the population. This is what is meant by natural selection.

An individual's probability of surviving and reproducing is called its fitness. Each individual will, in a given environment, have a certain probability of escaping death from predators, food shortage, parasites, disease and the weather. The phrase 'survival of the fittest' refers to this probability. Natural selection must be thought of as a statistical process in which certain individuals stand a better chance of surviving and leaving offspring than others.

Under most circumstances natural selection imparts genetic stability to a population: the same kinds of individuals are the survivors generation after generation. But if the environment alters, a different set of individuals with different genetic characteristics may stand a higher chance of surviving and reproducing. Given time, natural selection can alter some of the characteristics of the population so that different individuals emerge which are better adapted to the new environment. This kind of adaptive change is what is called evolution. Species are continually subjected to the possibility of alteration, but the process of evolution is usually slow and we would not, under normal circumstances, see new adaptations developing.

Natural selection usually acts on the existing variation among individuals of a population. But genes are also liable to mutations, which are potential sources of new variation in the population. Mutations occur infrequently, often no more than once in every ten thousand to a million individuals. They are haphazard and normally disadvantageous changes in the chemistry of genes with the result that affected individuals are quickly eliminated by natural selection. Occasionally, however, an advantageous mutation occurs and, if the environmental conditions are right, it quickly spreads through the population. For this to happen the environment has to favour the mutant individuals to an extent that their chances of survival are greater than non-mutants. The case of the peppered moth, to be discussed in the next section, is a good example of this.

There are two requirements if natural selection is to be important in maintaining or changing protective and deceptive adaptations in animals which are potential prey. The first is that predators are selective in the prey items they find or miss; the second is that the adaptations possessed by the prey are under genetic control of some sort and can be passed from parents to offspring. Both of these requirements are fulfilled in those species that have been intensively studied, and there is every reason to suppose that they are fulfilled in all species. Very often the inherited adaptation is a colour or pattern but it may also be a piece of specific behaviour, for example, the ability of a camouflaged moth to select a background of the right colour to rest on.

Some animals inherit the ability to change their colour or pattern. Thus certain African insects, among them the grasshopper, *Aulacobothrus wernerianus*, and the mantid, *Galepsus toganus*, turn black on burnt grassland but do not do so if there is no burning. The colour change reduces their chances of being seen by predators in an environment that has suddenly and dramatically altered. In these two insects the colour change is not reversible but in many other animals, particularly chameleons, certain fish, amphibians and cephalopods (octopuses, squids and cuttlefish) the capacity for reversible colour change is inherited.

The origin of a protective adaptation: the case of the peppered moth

The peppered moth *Biston betularia* is common in most parts of northern Europe and occurs eastwards to China. A similar species, *Biston cognataria*, is found in North America. The moths fly at night in early summer and rest motionless by day on tree trunks, but are rarely seen unless they are attracted to bright lights at night. The typical coloration of the wings is white or creamy-white, finely dotted and speckled with black. When at rest on a lichen-covered tree trunk they are beautifully camouflaged and difficult for predators to find.

In the first half of the nineteenth century virtually all peppered moths were of the typical pale form but a few

Pale and black forms of the peppered moth resting on lichen-covered bark. The pale one is beautifully camouflaged.

burning power stations which combines with rain to form weak sulphuric acid. In many towns and cities lichens have disappeared, but they still remain in unpolluted areas. Tree trunks without lichens tend to be uniformly dark brown or blackish, and they may be further darkened by the deposition of soot. Observations have shown that small insect-feeding birds searching tree trunks for food items find and eat peppered moths which do not match the background well. In cities these are the pale individuals, in the countryside the black ones. Experiments have also shown that black moths are capable of selecting a dark background, and pale moths a light background, on which to rest.

In the peppered moth natural selection by insect-eating birds has led to the spread of a new adaptation, but the story is a little more complicated than this. Some black moths appear to be hardier than pale individuals, possibly because their caterpillars are better able to cope with food-plants contaminated by pollutants. Moreover there is another dark form, under separate genetic control, which produces a moth intermediate between the typical and the all-black forms. This form is frequent in and around smaller towns where there is less pollution, but its frequency in areas of heavy industry is difficult to estimate because of the presence of the black form. Recently there has been a slight decline in the frequency of black peppered moths in parts of Britain where clean-air policies have resulted in less pollution.

The peppered moth's pale, intricately patterned protective coloration undoubtedly served it well, possibly for thousands of years, until the advent of the Industrial Revolution in the middle of the nineteenth century. We can assume that black individuals occasionally appeared as rare mutants but did not survive. Then as the environment changed pale moths were at a sudden disadvantage in certain areas and black moths were able to increase in frequency as a result of natural selection by birds.

had entirely black wings, the result of a mutation. By 1895 most of the peppered moths in and around industrial cities in the north of England were of the black form. There had been a sudden and dramatic change which was related to the growth of industry, and therefore of pollution, as the Industrial Revolution gathered momentum. Black moths did not increase in frequency in rural areas which remained free of pollution. Today in Britain, 98 per cent of the peppered months are black in areas of heavy industry, such as Manchester, while in rural areas, such as Cornwall, all the peppered moths are of the original pale form. Black peppered moths also occur at high frequency in and around many European cities and a parallel change, starting about 1900, occurred in *Biston cognataria* in North America.

Lichens growing on tree trunks are extremely sensitive to pollution, especially to sulphur dioxide from coal-

The two forms of the peppered moth on bark from which the lichens have disappeared as a result of industrial pollution. This time it is the black one that is more difficult to see.

In this example an environmental change precipitated by man resulted in the spread of a new protective adaptation.

The genetics of adaptation

Although the pale and black forms of the peppered moth look very different from each other they are controlled by a simple genetic system. This system, or some modification of it, controls alternative colour forms in many other species of animals and is known to occur in plants as well.

Genes are carried on chromosomes which occur in pairs in the cells of organisms. When the sex cells (in animals, the sperm and eggs) are formed, a special sort of cell division takes place in which the chromosome pairs separate. At this stage the sex cells contain half the number of chromosomes of the other cells in the body, and only one chromosome from each pair. Upon fertilization paired chromosomes are formed again, one set coming from the father and one set from the mother.

A gene occurs at a specified site on both paired chromosomes. At each site or locus, it is present in one of a number of alternative forms or alleles. In the black and pale forms of the peppered moth there are two alleles, but often there are more than two, sometimes as many as ten, as for example in certain mimetic swallowtail butterflies.

The letters B and b can be used as symbols for the alternative alleles associated with the black and pale forms of the peppered moth. Both BB and Bb individuals are black and are identical in appearance: they are of the same phenotype, a word used to describe what an organism looks like. But bb individuals are pale and are of a different phenotype. The word genotype is used to describe what an organism is in genetic terms, thus black BB and black Bb are the same phenotypes but different genotypes. When present with B the expression of b is masked and we see a black moth. When B expresses itself fully and suppresses the expression of b the phenotype is said to be dominant. When b is combined with b it gives what is called a recessive phenotype. In some organisms, including certain butterflies and moths, there are pairs of alleles which, when combined, give a phenotype intermediate in appearance; when this happens it is called incomplete dominance.

It follows from this that the kinds of offspring that appear from a mating depend on the genotypes of the parents. Different genotypes of parents give predictable ratios of phenotypes in the offspring. Thus a mating between a black BB and a pale bb gives only black Bb offspring. The consequences of a mating between two black Bb individuals are shown in the diagram on p. 00. The expected offspring from such a mating occur in the ratio $2\,Bb:1\,BB:1\,bb$, which means that if there were two hundred offspring there would, on average, be

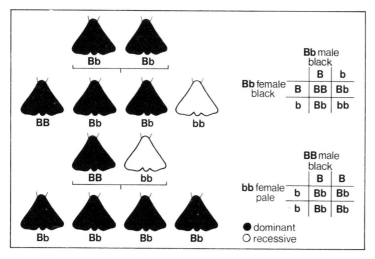

	Bb male black	
	B	b
Bb female black B	BB	Bb
b	Bb	bb

	BB male black	
	B	B
bb female pale b	Bb	Bb
b	Bb	Bb

● dominant
○ recessive

100 *Bb*, 50 *BB* and 50 *bb*.

Some colours and patterns are determined by the cumulative effects of many genes. This mode of inheritance is called polygenic, and is less easy to study than inheritance involving alleles in which the effects on the phenotype are clear-cut. Polygenic inheritance tends to produce continuous variation in colour and pattern (and in other characteristics) rather than well differentiated alternatives. If we take a sample of animals and find that the colours of individuals vary between, say, grey and brown, with most individuals intermediate between the two colours, there is good reason to suspect the existence of polygenic inheritance.

Polymorphism

The existence of distinct, genetically determined forms within a population, as seen in the peppered moth, is called polymorphism. The existence of two sexes is really polymorphism but is usually excluded as a special case. Polymorphism is discontinuous variation within a population, and under most circumstances polygenic inheritance does not lead to polymorphism. The complex pattern of a butterfly's wing might be expected to be determined by the cumulative effects of many genes, *ie* by polygenic inheritance, but in many polymorphic butterflies and moths the pattern and colour are in fact determined

by alleles at a single locus. There are generally a number of alternative alleles (called multiple alleles) and these allow a number of distinct forms to be produced. Intermediates between the forms may occur (perhaps produced by incomplete dominance), but at low frequency. Other genes, known as modifiers, may act to prevent the formation of intermediates.

The existence of polymorphism reflects a balance of selective forces such that under certain conditions in the environment one form is at an advantage in terms of survival while under other conditions another form is at an advantage. The frequency of different forms in a population may change, as in the case of the peppered moth, but often a period of change is followed by stability at a new frequency; on the whole polymorphic forms tend to remain at about the same frequency generation after generation.

The continuing need for deception

In this chapter various components of adaptation are recognized. Natural selection is the great unifying principle: an understanding of it in a genetic context helps to explain the evolution of deceptive and protective adaptations in both predators and prey. There are many variations on the theme which we shall explore in the following chapters, paying particular attention to polymorphic species which provide the best evidence for the effectiveness of protective strategies. The evolutionary conflict continues and we are privileged to see and investigate a remarkable array of ploys and counter-ploys among both the hunters and the hunted.

3   A green and brown world

The main colours of vegetation are the greens of living leaves and young stems and the browns of twigs, branches, bark and dead leaves. The extent of vegetational cover varies markedly in different parts of the world, from none or very little in arid deserts, to a dense stand of tall evergreen trees in tropical rain forests. Everywhere there are seasonal changes. In the temperate regions grassland changes abruptly from green to brown with the onset of autumn and the leaves of trees turn colour before dropping to the ground. In many temperate areas, particularly eastern North America, the reds and yellows of autumn foliage are a temporary exception to the dominance of summer greenery and winter drabness.

In the humid tropics many species of trees produce new leaves at the beginning of the rainy season but, unlike the new leaves of most temperate region trees, they are often red or scarlet: a visitor to a tropical forest could be forgiven for mistaking what he sees for autumn colours. During the dry season the grass of tropical savanna is dead and brown; it looks like standing hay, and only for brief periods when there is plenty of rain is the environment conspicuously green. The monotony of green and brown is broken here and there by the bright colours of flowers. But, except in gardens where the abundance of flowers is artificially high, the contribution of flower colours to the overall scene is usually small: wherever there is vegetation greens and browns predominate.

Many small animals have evolved green or brown camouflage, and a remarkable number of species, especially insects, have green and brown colour forms. The resemblance to leaves, twigs and bark is particularly close in species that are active at night and remain motionless by day in exposed places. The resemblance is less close in the many species that spend the day in hiding.

Many large animals, including predators like lions are also brown and harmonize well with their background. Almost all mammals are countershaded: that is, they are some shade of brown on their backs, while the fur on their undersides is much paler, often white. In some the brown coloration is broken by pale or dark stripes. The okapi of African rain forests has striking disruptive coloration of this kind at its rear end; as it runs away from danger and disappears into the forest its outline is well broken up. The stripes of the tiger look as if they might act as disruptive coloration also; possibly they do so, and aid the tiger in stalking prey, but this would be very difficult to demonstrate experimentally. The young (but not the adults) of some large mammals, including several species of deer, are spotted which presumably helps to conceal them from

Left This South
American sloth looks
green but in fact the
greenness is caused
by small algae which
grow on the hair.

Right The Eurasian
bittern nests among
reeds. When alarmed
it stretches its neck
and sways its body
with the reeds and,
because of its striped
pattern, becomes
almost impossible to
see.

The tropical rain forest is a varied and patchy environment. The great potoo of South and Central America sits still on branches and resembles a piece of dead wood and, unless it moves, fits well into the jigsaw of vegetation.

predators. Spotted markings tend to have a concealing effect in forests, where sunlight falling through the leaves makes a dappled pattern of light and shade on the undergrowth. But whether the distinctive coloration of such species as the giraffe and the zebra are examples of camouflage is less certain.

Except in birds, snakes and a few lizards, green is an uncommon colour in large animals. There is no truly green mammal, although green algae grow on the hairs of the slow-moving, tree-living sloths of the rain forests of Central and South America. The algae give a greenish appearance which may help camouflage the sloths in their leafy surroundings. The green is best developed in the rainy season when the algae thrive; in the dry season the algae die back somewhat and the sloths look greyish-brown. But monkeys, many of which are foliage-dwellers, are not green, and presumably rely on their size and agility as a means of escaping predators. It is possible that mammals are simply unable to produce a green coloration in their fur.

Birds which live among foliage are often greenish above and pale below, another example of countershading. Some species of camouflaged birds, such as the chiff-chaff and willow warbler of Europe, are so similar to each other that they cannot easily be identified except by their distinctive songs. Ground-dwelling birds are often brown with intricate dark markings: a nightjar or a female pheasant sitting still among dead leaves is almost impossible to see. A few birds are camouflaged to look like dead branches when sitting in trees. One such is the great potoo

Nyctibius grandis of South and Central America. A drab greyish-brown, marbled with black, it accentuates its resemblance to dead wood by freezing with its head pointing skywards when disturbed. The head and beak look like the sharp, splintered projection of a broken-off branch. Very similar behaviour is seen in the Eurasian bittern *Botaurus stellaris* which nests among reeds. When alarmed it too stretches its head up in a straight line and sways with the reeds as they are blown by the wind. Vertical stripes on its body which continue up the neck to the head, blend well with the background provided by the thickly growing reeds.

Not all birds are camouflaged; some are brightly coloured and boldly patterned, the colours and patterns functioning in courtship and display and defence of territory. The males of many species are much more brightly-coloured than the females, presumably because the females are more vulnerable to predators as they incubate eggs and young and because the male makes use of bright and bold plumage coloration in courtship and defence of territory. In many other birds – the American wood-warblers are a good example – bright male plumage is developed only in the breeding season; throughout the rest of the year they are rather drab looking. The idea that females are camouflaged because they are the ones which incubate the eggs is borne out by species in which these roles are reversed. There are a number of species of birds in which the male sits on the nest. One of these is the red-necked phalarope which breeds on lakes and pools in the far north of Europe and Asia. Here the summer season is very short, and the terrain offers little cover from predators. Probably as an adaptation to this, it is the male phalarope which builds a nest and later incubates the eggs. The female phalarope, after laying her eggs, moves on to another male's nest and begins again. In this way, at least one of her broods may survive. In most birds,

if a predator destroys the eggs or young, the female will begin again, but, because summer is so short in the far north, the red-necked phalarope would be unable to do this. What is interesting about the red-necked phalarope is that it is the female which has bright plumage, while the male is both smaller and duller in appearance. This is true of other species where the male incubates the eggs, and suggests strongly that the bird which sits on the nest is camouflaged.

In birds, as in most groups of animals, colours and patterns result from a balance of selective forces: in some species camouflage is all-important, in others it is non-existent; matching the green and brown world is necessary for some but not for all.

Green and brown grasshoppers and mantids

If a grasshopper matches beautifully the green grass upon which it rests it is all too easy to assume that it is camouflaged and that its colour is deceptive to potential predators. But this is a subjective judgement, where we are, in effect, putting ourselves in the position of a predator. We are not really justified in doing this, for no matter how well an animal may appear camouflaged to us we cannot be sure that a predator sees it in the same way. The power of vision varies immensely from one group of animals to another, as described in Chapter 1. Some animals lack man's faculty of colour vision, while others can see ultraviolet wavelengths which are invisible to us.

How can we prove that what seems to be camouflaged is, in fact, camouflage? We need to show that predators take animals which are not camouflaged more readily than camouflaged ones. One difficulty in trying to do this is that if all individuals of a species are similar in colour and pattern there is no easy way of showing experimentally that predators are selective; slight differences between individuals may determine which survive and which die

Overleaf Many small animals are adapted to the green and brown world by being either green or brown. Although these two mantids are differently coloured, they belong to the same species. The green form is usually found in living vegetation while the brown form is more often found in dead vegetation.

A multiplicity of colour forms in the African grasshopper *Ruspolia differens*. The various combinations of green, brown and purple match the colours of the grasses in which the grasshoppers live.
1 green
2 purple-striped green
3 purple-headed green
4 brown
5 purple-headed brown
6 purple-striped brown

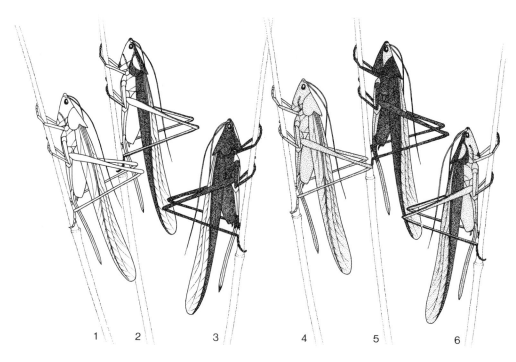

1   2   3   4   5   6

but to grade these differences and to show that predators respond to them is virtually impossible. Instead, for experimental purposes, it is necessary to investigate species in which there are clear-cut alternative colour forms, that is to say, polymorphic species.

Many grasshoppers living in patchy green and brown environments have green and brown colour forms which are either under direct genetic control or are produced in response to stimuli from the environment in which they occur. If given a choice individuals often select the appropriate background colour which alone suggests that the colour forms are involved in the deception of predators. The African grasshopper or bush-cricket, *Ruspolia differens*, is abundant and in some localities occurs in enormous swarms during the rainy season. It is about 6 centimetres long, which is large for a grasshopper, and either green or brown; some individuals are also marked with purple. The adult grasshoppers are active at night and remain hidden in grass during the day; they are capable of long-distance migrations, apparently triggered by the alternation of wet and dry seasons. In East

Africa they are much prized as food. On wet nights thousands are collected from beneath street lights to which they are attracted. Because they are edible and regarded as a delicacy, a considerable folklore has developed around them with all sorts of explanations as to where they come from and, interestingly enough, why they occur in several different colour forms.

A sample of 10,549 *Ruspolia differens* collected at a single site in Uganda contained the following colour forms: 6682 green, 3521 brown, 305 purple-striped green, 34 purple-headed green, 5 purple-headed brown and 2 purple-striped brown. In this sample green is almost twice as frequent as brown, and together these two forms account for about 97 per cent of the total. The remaining four forms are relatively rare, together comprising about 3 per cent. The purple-striped brown form occurs at a frequency of about one in five thousand; if only a single individual of this form had been found in the sample it might have been regarded as an unusual mutant and not a form maintained in the population by natural selection.

The three conspicuous colours in the

habitats of the grasshoppers are the same colours that occur in the grasshoppers themselves; the colours and combinations of colours can be similarly ranked in the habitat of the grasshoppers, suggesting that each of the forms is adapted not only to grass colour but also to the amount of each colour. The green form matches living grass exactly; likewise the brown matches dead grass. The purple of the remaining four forms resembles the purple streaks which often occur on grass leaves and stems. In the habitats where the grasshoppers live, green is the commonest colour, brown is easily next, and the purple on green or brown leaves and stems is the rarest. There is a higher frequency of greens in females than in males which can be accounted for because females spend much time among living grass while egg-laying. At least sixteen species of birds including kites, herons, storks, cranes, kingfishers, shrikes, crows, starlings, sparrows and chickens feed on *Ruspolia differens*, and it has been suggested that the annual migration of Abdim's stork follows that of the grasshoppers. Mammals also eat them; among the species recorded are rats, jackals, civets, genets, mongooses, dogs and cats. This array of predators and the remarkable resemblance between grasshopper and grass colours indicates that the polymorphism has been evolved as a means of camouflage where the background colour is variable.

Experiments have been conducted on other species of grasshoppers to show that their colours do act as camouflage and protect them from predators. These experiments involve placing about equal numbers of green and brown individuals among green or brown vegetation. Grasshoppers on the 'wrong' background are quickly seen and eaten by predators such as birds and chameleons, thus providing evidence of the effectiveness of the camouflage and the polymorphism.

Several species of mantids (which are related to grasshoppers) also occur in green and brown colour forms. In an experiment first conducted in 1904 and since repeated, green and brown forms of the European praying mantid *Mantis religiosa* were tethered to green or brown vegetation, left for some days, and the survivors counted. The greens disappeared from brown vegetation and the browns from green vegetation. Insect-eating birds were responsible and they were quick to select individuals that did not match the background.

In this experiment the mantids were forcibly confined; in nature they are able to move around and select a background, a useful attribute if green and brown forms are to be effective as a means of camouflage. In several species of grasshoppers and mantids green forms are better at selecting green backgrounds than brown forms are at selecting brown backgrounds. Precisely why this is so is not clear but possibly green on brown is more conspicuous than brown on green; perhaps in nature it is more common to find a dead leaf among living leaves than a living leaf among dead ones.

Green and brown butterfly pupae

The pupae of most species of butterflies are attached to living or dead vegetation and appear well camouflaged. Colours and patterns vary with the species: many are green, brown, or grey; some are speckled or dotted with black or dark brown, others are unmarked and a few have glistening metallic markings. Butterfly pupae are difficult to find. The European orange-tip *Anthocharis cardamines* remains in the pupal stage from August to May, a long time to be motionless and defenceless on the stem of a plant. The pupa is extremely well camouflaged, green at first, then turning pale grey or light brown; it is finely tapered at each end, and bent in the middle, with bulging wing-cases. It looks exactly like a piece of bent stem. To find one is something to talk about.

Pupae of the European swallowtail *Papilio machaon* vary in colour between

Camouflaged butterfly pupae: **1** the green and brown pupae of the small white; **2** the green pupa of the orange tip which often turns brown; **3** the well-positioned and remarkably leaf-like pupa of the African swallowtail, *Papilio dardanus;* **4** instead of looking like a leaf, the pupa of an Asiatic swallowtail looks like a piece of snapped twig.

green and black but most can be classified as either green or brown. The pupae are attached to reeds; a silken pad secures them at the rear end and support is provided by a silk girdle around the body. Pupae resulting from butterflies flying in May and June are often green and match green reeds, but some are brown. These green pupae usually produce a second generation of butterflies in August. Overwintering pupae are derived from part of the first and all of the second generation of butterflies, and nearly all of them are brown. Thus in this species the frequency of greens and browns depends on the time of year and on whether or not they hibernate. In an experiment on swallowtails conducted in Sweden it was found that in summer green pupae survive better on green backgrounds and brown pupae on brown backgrounds, but in winter the survival rate is less dependent on background colour. In summer birds are the predators, but in winter it is mainly shrews and mice. Birds have colour vision while shrews and mice do not, which seems to explain why green and brown pupae which match the background are advantageous in summer but not in winter.

In West Africa about seventeen species of swallowtails may occur at a single locality. In one species, *Papilio dardanus*, the pupa is attached to the food-plant of the caterpillar by a silk girdle to a leaf and the rear end to a stem; it is flattened, invariably green, and remarkably leaf-like. Ridges simulate the veins of a leaf, and the pupa is positioned halfway between two other leaves, in such a way as to make it best look like a leaf of the plant. In another species, *Papilio menestheus*, the pupae are attached to stems and bark; they are light brown, strongly marked with dark brown, and with just a touch of green. They resemble a piece of stem or bark with a slight growth of green algae. In most (possibly all) of the remaining fifteen species the pupae are either green or brown, like those of *Papilio machaon*.

Green pupae are usually attached to green leaves and brown pupae to dead leaves or stems. One species, *Papilio demodocus*, produces more brown pupae in the dry season and more greens in the wet, a seasonal switch which makes sense because in the dry season the land is parched and brown while in the wet it is lush and green.

The pupae of the small white butterfly *Pieris rapae* are similarly green or brown. Nowadays the caterpillars of this species feed chiefly on cultivated cabbages; at one time they must have fed on wild species of Cruciferae, the plant family to which cabbage belongs. Pupae attached to leaves are usually green while those on dead vegetation, walls and fences are usually brown. There are two, sometimes three, generations of small whites in a year; pupae resulting from late summer butterflies overwinter and are usually brown; green pupae are most frequent in summer and are more often attached to leaves than to anything else.

The existence of green and brown colour forms in the pupae of butterflies leads to two questions: how is it done, and why? The second question is easily answered: it is a response to a patchy, often seasonally variable, green and brown environment; the pupae are well camouflaged and, since they do not move, are often overlooked by predators. The first question is more difficult. Clearly we are not dealing with an example of polymorphism in the strict sense defined in Chapter 2 because the two forms are not under direct genetic control: it is impossible to predict the ratio of greens and browns from a given mating, even if the colours of the parents' pupae are known. What appears to be inherited is the ability of the fully-grown caterpillars to respond to conditions in the environment selected for pupation and to turn into green pupae or brown pupae as appropriate. There is no way of discovering how this is done by field observation and so the question must be answered from the results of laboratory experiments.

The advantage of laboratory experiments is that the environment can be controlled and measured with a good deal of precision. By rearing caterpillars in different environments it is possible to discover which factor or factors control pupal colour. In the small white butterfly light is the most important factor: exposure to yellow light usually results in green pupae, while blue light results in brown pupae. This result fits in with what happens in nature: light falling through green leaves contains an excess of yellow wavelengths; blue wavelengths are more pronounced when there is no green foliage. In the tropical Asian swallowtail *Papilio polytes* a rough texture to the substrate on which pupation occurs results in more brown pupae and a smooth texture more green pupae. Again this fits in with what occurs in nature because this swallowtail, unlike most others, pupates at night and could not therefore respond to light wavelengths. However, it is not known what the rough and smooth textures correspond to in nature. Most species studied in the laboratory also respond to temperature, to duration of light and dark periods, and to humidity. High humidity often results in a high frequency of green pupae and low humidity in more browns, another result which fits in with the natural environment, especially in the tropics, where during the wet season the air is more humid and there are more green leaves.

What is interesting is that no two species respond in exactly the same way, and even within a species there are variations associated with the geographical area from which the experimental stock originated. Varying the background colour by offering caterpillars differently coloured tissue paper on which to pupate also affects the frequency of green and brown pupae. In one experiment involving caterpillars of the small white butterfly, brown pupae were produced on a background of red, blue or white tissue paper, and green pupae on orange or yellow paper. A result like this shows that background colour affects pupal colour even though the colours used in the experiment would not be expected to occur where pupae are formed in the wild.

The ability of a pupating caterpillar to respond to these stimuli depends on the presence of a hormone believed to be secreted by the brain. There is thus an interaction between external conditions and internal physiology, a conclusion also reached in laboratory experiments which were designed to answer the same sort of 'how' questions about the production of green and brown colour forms in young grasshoppers. Interestingly, in these laboratory experiments, it proved possible to produce brown forms of young grasshoppers which in the wild have only green forms.

Laboratory experiments do not of course explain how a caterpillar selects either a green or a brown background on which to pupate. Possibly there is no selection as such: the caterpillar simply locates a suitable site and is then programmed to produce a green or a brown pupa according to the colour or texture of that site and in response to subtle differences in light, humidity and temperature in its immediate environment.

All of these physiological and behavioural responses have been evolved and are maintained by natural selection exerted by predators, especially by insect-eating birds. Camouflage in pupae reduces the risk of discovery; it is an example of what is called primary defence. Camouflage is of paramount importance to pupae, for, once discovered, and recognized as potential food there is nothing a pupa can do to escape because it is tied down securely with silk. It cannot even drop and disappear into dense vegetation in the manner of many caterpillars.

Camouflaged caterpillars

Caterpillars are usually green if they spend most of the time on the leaves they eat, or some shade of brown if they rest by day on twigs and feed on leaves at night. The colours and patterns are under genetic control but there are many species in which stimuli from the environment alter the colours so that there is an even better match of the background. Green and brown colour forms occur in a wide variety of species just as they do in grasshoppers, mantids, and butterfly pupae.

Some green caterpillars, including those of the silver-Y moth *Autographa gamma* and the elephant hawk-moth *Deilephila elpenor*, tend to turn dark when crowded. During its lifetime an elephant hawk caterpillar moults its cuticle five times. The caterpillars are normally green until after the fourth moult when almost all become blackish-brown; a few remain green, and a few turn dark after the third moult. Caterpillars reared singly in glass jars remain green after the third moult, but it reared in groups of five or more many turn dark. Crowded caterpillars also mature more quickly and pupate a few days earlier than those reared singly. Is this density-dependent colour change related in any way to camouflage? Nobody knows, and it is certainly difficult to imagine how it might give protection from predators. We must remember that the colours of animals like caterpillars may have other functions. Thus the dark coloration developed by crowded elephant hawk caterpillars might be a form of communication between individuals competing for food — crowded caterpillars certainly feed and grow more quickly than solitary individuals. There is need for investigation here, and also for caution: we must not assume that the colours of caterpillars are invariably associated with protection from predators.

On the other hand the twig-like caterpillars of moths of the family Geometridae are notorious for their camouflage and it is difficult to imagine that their coloration has any other function. This family of moths is found throughout the world and in most places

All the features of a twig are shown in this geometrid caterpillar. Provided it keeps still, there is little chance of it being discovered by a predator.

there are many species some of which are abundant. Resemblance to twigs is achieved by a combination of colour, which closely matches the background on which the caterpillar is found, the presence of humps and marks on the body which look like buds and scars on twigs, the absence of abdominal legs except the hind pair which the caterpillar uses to grasp the twig, and the caterpillar's habit of resting in an unusual position with the body stretched at an angle from the branch. Many obtain support for this uncomfortable looking position by spinning a silk thread between the head and a leaf or twig just above it. Some species adopt a humped posture and so look like a bent twig.

Each species resembles the twigs of its own food-plant, and if placed on a different plant the effectiveness of the camouflage is much reduced. The caterpillars of the European brimstone moth *Opisthograptis luteolata* normally feed on the leaves of hawthorn, blackthorn, and similar trees. They are greenish to brownish and match the twigs of these trees well. In Britain the brimstone moth has started utilizing the leaves of the myrobalan plum *Prunus cerasifera* which was introduced from Persia in 1880 and now grows commonly in gardens as an ornamental tree. The leaves and young twigs are reddish-purple and caterpillars found on them occur in two forms: green and dark reddish-purple. Although twig-like, green caterpillars are conspicuous, but reddish-purple individuals match beautifully the purple foliage and twigs. None of the native food-plants of the brimstone has purple leaves or twigs and so it seems likely that the purple colour form of the caterpillar is a recent development. If a brood of brimstone caterpillars is divided and half reared on green and half on purple leaves, most of those on green leaves become green while most of those on purple leaves become purple. This suggests that the colours are not under direct genetic control but are developed in response to the immediate surroundings in much the same way as green and brown butterfly pupae. There is some evidence with grasshoppers that red pigments in their food can be absorbed and can affect the colour of the insect. Whether something

Right A caterpillar of the European brimstone moth. The position in which it rests and the bud-like lump on its back enhance its resemblance to a twig.

Opposite page These two forms of the caterpillar of the African swallowtail, *Papilio demodocus*, are controlled by a single gene difference. The top photo is the form normally found on citrus plants while the bottom photo is that normally found on umbelliferous plants. Each form survives best on its normal food-plant because it is less easy for predators to find.

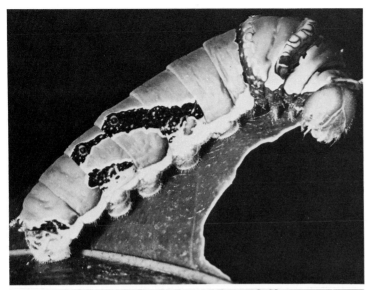

similar might be happening with brimstone caterpillars is not known.

Experiments and observations show that birds take a long time to find the twig-like caterpillars of the Geometridae. In one experiment European jays were offered real twigs and caterpillars of various species. If offered twigs of the caterpillar's normal food-plant with a few caterpillars mixed with them the jays experienced particular difficulty in distinguishing the two and even when hungry often gave up. But if offered twigs from another species of tree mixed with caterpillars the jays were more successful. The jays pecked at twigs and if by chance they discovered a caterpillar they then searched for more, but soon became discouraged if they found nothing but twigs. The experiment demonstrates that predators are quick to learn from a successful experience, but soon give up a search that yields nothing.

Caterpillars of the Geometridae, perhaps more than any other group of animals, combine coloration, stance and behaviour most effectively in their resemblance to twigs of the food-plant. A predator cannot try every twig to determine whether it is really a caterpillar, yet seemingly this is what they have to do. How do predators ever find geometrid caterpillars? The most likely answer is that individual caterpillars give themselves away by slight movements which are quickly detected and investigated by predators; it is also possible that although these caterpillars look to us exactly like twigs, some predators have a different view and find them more easily.

The swallowtail butterfly, *Papilio demodocus*, occurs throughout Africa south of the Sahara; its caterpillars feed mainly on the leaves of trees of the family Rutaceae, many species of which occur in both forest and savanna. Almost everywhere there has been a switch to orange, lemon, lime and grapefruit trees which have been introduced by man. These trees are all of the genus *Citrus* which also belongs to the

Rutaceae. Fully-grown caterpillars are green with pale-edged darker markings which look like leaf blemishes, and they appear well camouflaged on *Citrus* leaves. In parts of South Africa the caterpillars also feed on the leaves of plants in the family Umbelliferae, which includes fennel and parsley. The leaves of these plants are fern-like or feathery and quite unlike the leaves of *Citrus*. Many (but not all) of the caterpillars on Umbelliferae are markedly different in appearance from those on *Citrus*. They are green, but intricately marked with brown, so that they are well camouflaged when on Umbelliferae, but not when on *Citrus*. It was once thought that the food-plant determined which kind of camouflage developed but this is not the case. The two forms are determined genetically, so that the caterpillar is destined to be of the citrus-type or umbellifer-type before it hatches from the egg. They are controlled by a pair of

In England, the large wainscot moth sits about on dead reeds.

alleles with the umbellifer-type dominant to the citrus-type. Citrus-type caterpillars are most often found on *Citrus* and umbellifer-type on Umbelliferae, and experiments confirm that birds are better at finding those on the 'wrong' background. What is not known is how the egg-laying female butterfly chooses the correct kind of food-plant for her caterpillars, if indeed she does choose. Another unanswered question is why the species has two very different kinds of food-plant in only a small part of its range.

Papilio demodocus has successfully exploited cultivated *Citrus* throughout Africa south of the Sahara and is now an abundant butterfly in gardens and farmland, where its caterpillars are occasionally pests. In West Africa caterpillars have been found feeding on marigolds and related plants of the family Compositae. Some of these have leaves rather like the Umbelliferae and it will

Above A membracid bug from Trinidad looks just like a chewed leaf.

Right Looking like a leaf is a common protective strategy. This bush-cricket (katydid) lives in the rain forests of Costa Rica.

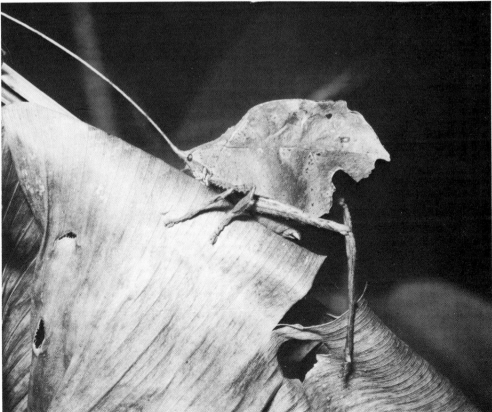

be interesting to see if the umbellifer-
type caterpillars begin to appear outside
their present range in South Africa.

Leaf-like animals

In many animals camouflage against a
background of vegetation has developed
into a camouflage in which the animal
actually resembles a leaf, as in the pupa
of *Papilio dardanus* already described.
Such leaf-resemblance is best developed
in insects, which have the advantage of
already having veins in their wings.
These can be rearranged and accentu-
ated by the evolutionary process to
simulate the veins of a leaf, as is seen in
some of the bush crickets and mantids.
These resemble green leaves almost perfectly: the disguise even includes small
brown marks which look like damage
by leaf-eating insects.

In certain butterflies, such as the
Kallima species of Africa and Asia, the
leaf veins are not represented by the
butterflies' own wing veins, but are

A South American fish, *Monocirrhus polyacanthus*,
and the leaf it resembles. Even the barbel on its
chin looks like a leaf stalk.

'painted on' as part of the pattern produced by tiny scales on all butterflies'
wings. This involves long, narrow
markings on both fore- and hindwings
so positioned as to form a single straight
line when the butterfly holds it wings
together at rest. The *Kallima* species
are brown to resemble dead leaves and
this is also true of the Eurasian lappet
moth which, however, looks like a

Some noctuid moths sit by day on twigs. They cling on in such a way that they appear to be part of their background.

which look like partially eaten leaves.

Vertebrate animals can also resemble leaves. Several species of frogs and toads which live on the forest floor are strikingly leaf-like and extremely difficult to pick out from their background. In the horned frog *Megophrys nasuta* of Malaya, the body is flattened, with sharp outer edges and triangular horns over the eyes so that these are hidden. The frog is patchily coloured brown and has a line down its back looking like the midrib of a leaf.

Even some fish resemble leaves, float on the surface of the water. The leaf-fish, *Monocirrhus polyacanthus*, from the basin of the river Amazon in South America, is mottled brown and flattened sideways. As it drifts head downwards near the surface of the water, it looks remarkably like a floating leaf and, using this cover, it seizes small fishes in its protrusible jaws with lightning rapidity.

Selecting the right background

When a swallowtail caterpillar is ready for pupation it selects a leaf or twig, attaches itself, and then moults into the pupal stage. The new pupa is soft and delicate but quickly hardens and develops its colour. A pupa cannot move or change colour: it must remain for weeks or months exactly where it was formed. Selecting the right background is thus vitally important: what the caterpillar does influences whether its pupa escapes being found by a predator.

Mobile insects like caterpillars and moths also select the appropriate background on which to rest but have the advantage of being able to change position. Many moths rest by day on tree trunks. After a night's activity a moth selects a daytime resting place on the bark of a tree but rarely returns to the same place a second time. In most species of bark-resting moths the forewings are held over and obscure the hindwings. The forewings are beautifully camouflaged and either blend with the bark or give the moth the appear-

bunch of curled, dead leaves.

Some of the best examples of this sort of camouflage are found among the leaf insects, which belong to the Phasmida along with the stick insects, or walking-sticks as they are sometimes known. Stick insects, of course, are camouflaged as thin, branched twigs or as grass. In contrast, the bodies of the leaf insects are broad and flat, with veins and many other detailed resemblances to leaves. Their limbs also have irregularly shaped, papery outgrowths

ance of a dead leaf, piece of lichen, or some other object that might be attached to a tree trunk.

The most spectacular bark-resting moths are the underwings of the genus *Catocala*. One species, the red underwing *Catocala nupta*, is common in Britain and two others, the dark crimson underwing *Catocala sponsa* and the light crimson underwing *Catocala promissa*, are relatively rare, but in the woods of eastern North America thirty or more species occur at a single locality. The North American species are described in more detail on p. 88. The forewings are brown, grey, or whitish, intricately patterned with darker or lighter markings, and when the moths are at rest completely obscure the hindwings which are generally strikingly coloured as their names imply. Apart from their ability to

select the right background on which to rest, the moths orient themselves and attain a position which aligns their markings with those of the bark; some species even rest in an upside down position.

In the United States experiments have been conducted to test the ability of underwing moths to select the right background. Moths were released in a chamber painted with black and white bands which provided a choice of background differing only in reflectance. In the morning the resting positions were scored and it was found that species with pale forewings usually selected a white background and those with dark forewings a black background. In one trial 10 *Catocala relicta*, a species with pale grey, black-flecked forewings, were released into the chamber and all selec-

It is not good enough simply to resemble a background: the right background for resting must be chosen. The North American underwing moth, *Catocala relicta*, is able to select the background appropriate to its own wing coloration.

44

Sitting sideways often makes a better match. The European small blood vein moth at rest on a tree trunk.

Opposite page The Malaysian flying gecko flattened against the bark of a tree.

ted the white background. But of 66 *Catocala antinympha*, a species with dark forewings, 56 selected a black and 10 a white background. Thus as usual there were 'mistakes' but, of course, providing a simple choice between black and white is not the same as selecting the right background on something as complex and variable as the bark of a tree. How the moths achieve a good match between their own forewings and the colour and pattern of the bark is uncertain. One theory is that a moth can match its own reflectance with that of the background; another is that the behaviour is under genetic control and that the moth's colour and its behaviour when selecting a resting site are controlled by the same gene.

The bark of trees varies in colour between species but is predominantly some shade of brown or greyish-brown except when covered by a growth of greenish lichens or algae. Indeed in the green and brown world of a woodland in summer, tree trunks are the main source of brown backgrounds. At least 14 species of North American underwing moths have recently evolved black or blackish forewings, especially in areas where lichens have disappeared from bark as a result of pollution. But *Catocala relicta* has apparently always been variable with a continuous range of colour from pale grey to almost entirely black forewings; this can be explained by the patchy coloration of birch bark on which the moths normally rest.

The elimination of shadow

An important part of the camouflage of bark-resting moths is the elimination of the shadow cast by the body. This is achieved by holding the wings flat so that their edges touch against the bark as much as possible. Other animals which rest on bark have special extensions to the body which can be held against the bark and eliminate shadow in the same way. The flying gecko *Ptychozoon kuhli* of Malaysia has a wide flap of skin around the edges of its body

and tail, which curls underneath the animal when it is at rest. The name 'flying gecko' was given to this animal because it was thought that the flap of skin aided it in gliding from tree to tree, but further observation suggests that its chief function is in camouflage. Other geckos have been found in Australia and the West Indies with similar skin-flaps, and these too are well camouflaged against bark.

Underwater camouflage

The plants of the open ocean are mostly tiny green plankton living near the surface of the water; they lack the complex structure of land plants and do not provide a suitable background against which camouflage could be evolved. Small animals, especially immature crustaceans, molluscs and fish, feed on the green plankton. They are mostly transparent which makes them difficult to see in clear water.

Since light penetrates only a relatively short distance in water, plant life is effectively absent more than 30 metres down. Elaborate camouflage of the sort common in land animals therefore occurs mostly in species that live in shallow waters. In shallow parts of the sea, and in some areas near the surface of deep water, there are seaweeds which are often bulky and some shade of brown. Many animals blend well with the colour of seaweeds. Some species of sea-horse, notably the Australian sea dragon *Phyllopteryx eques*, as well as being the same colour as the seaweed, have long ribbons of skin growing from the body; these are branched and tattered so that they resemble the fronds of seaweeds. Similar outgrowths of skin, which enhance camouflage by breaking up the body outline, are found in the sargassum fish *Histrio histrio*. This fish lives in clumps of *Sargassum* weed which float on the surface of the ocean in many parts of the world; its patchy brown coloration and many wisps of skin all over the body ensure that it blends well with the weed.

In general the kinds of camouflage found in aquatic animals are similar to those of land animals but there are several which are better developed in water than on land. One of them is countershading. Virtually all of the thousands of species of fish that are not bottom-dwellers are silvery below and dark above. This a very elaborate form of countershading: the underside is not just paler, as in land animals, but actually reflects light from its silvery scales. Such elaborate countershading is needed underwater because light comes from one direction, and one direction only; on land most light comes from above, but there is also a fair amount of light coming from the side, and a little reflected up from the ground. When viewed from above countershaded fish blend well with the darkness of the depths or with plants, pebbles and mud. Although some of their predators, such as herons and kingfishers, strike from

above, most are other fish which live in the same depth of water. A fish swimming in open water cannot in the strict sense be camouflaged – there is no structural background – but its shape is obscured by the effects of countershading.

One species of freshwater fish, the Nile catfish *Synodontis batensoda*, has the unusual habit of swimming upside down. It does this in order to browse on plants growing on the water's surface. Interestingly, this fish shows reverse countershading: its back is lighter in colour than its belly. Reverse countershading is also found in water boatmen, freshwater insects which swim upside down. These instances of reverse countershading emphasize the importance of countershading to aquatic animals.

Although a few land animals, for example chameleons, are able to undergo rapid and reversible colour change in

Among floating sargassum seaweed, there are many well-camouflaged animals. This one is the sargassum fish *Histrio histrio.*

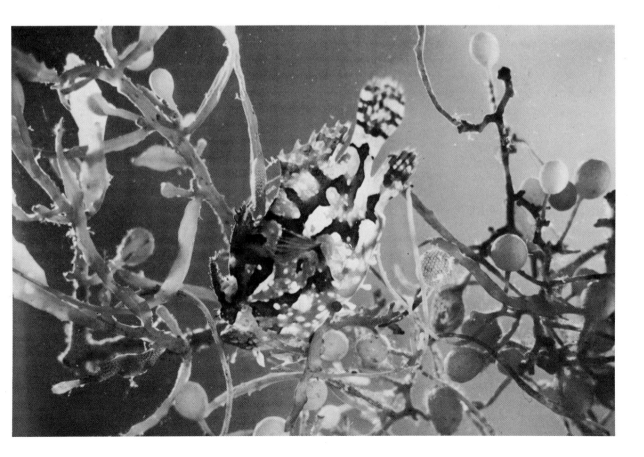

Opposite page
Always difficult to
see, chameleons can,
in addition, quickly
change their colour.
Even when found
together, no two
individuals are
coloured alike.

response to changes in their surroundings, most of the species that are able to perform this feat are aquatic and live on or near the bottom of shallow water. A variety of shrimps, prawns, octopuses, cuttlefish, bottom-living flatfish, freshwater fish, skates, rays, frogs, toads and newts possess the ability to change colour rapidly.

In frogs and toads and in many fish there are special cells in the skin which contain black melanin granules. These cells are intricately branched and the melanin granules may either be distributed throughout the branches or concentrated into a tiny speck within the cell. When dispersed the melanin overlays any other colour the animal may have and there is an overall darkening in coloration; when the melanin is concentrated the animal is pale. All kinds of intermediate shades can be achieved. In cuttlefish the pigment-containing cells occur in three layers with differently coloured pigments in each. The cells can be expanded or contracted by radiating strands of muscle under the control of the nervous system. As a cuttlefish moves to different backgrounds its colour and pattern changes almost instantaneously enabling a most effective harmonization with varying surroundings. In freshwater fish, colour change in response to a new background is rapid, but a perfect adjustment may

Woodlouse-like sea-
slaters can turn dark
or pale in response
to changes in light
intensity.

take several days. Experiments have been conducted in which fish were kept in black- or white-sided tanks and left to adjust their colour or tone. They were then released in equal numbers into tanks of the same colour or the 'wrong' colour. As might be expected, predatory birds and fish caught more fish that initially contrasted with the background than those that did not, but when left long enough all the fish adjusted to the new background. Fishermen who use minnows as bait for larger fish know the value of keeping them in white containers. The minnows become pale and, when first used, show up well in the murky depths of rivers and lakes and hence make better bait.

Plants that match other plants

It is not easy to imagine a camouflaged tree or bush and the suggestion that any plant might be camouflaged could be dismissed as improbable: after all, vegetation provides the main background for the evolution of green and brown camouflage in animals. There are, however, a few plants which blend with dead vegetation and by doing so are probably overlooked by grazing and browsing animals that locate suitable food-plants by sight. Plant-feeding insects tend to locate their food-plants by smell and are unlikely to be important in the evolution of camouflage in plants: possible exceptions are certain species of butterflies which use visual cues to find plants on which to lay eggs. Virtually all camouflaged plants are succulents or parasitic mistletoes growing in arid regions where browsing mammals occur. It is possible that there are many more examples than at present known: the difficulty is in finding the plants in barren, arid surroundings and in appreciating the subtleties of their coloration and growth form.

Several species of cactus have long, ribbon-like spines which resemble the shape and coloration of dead grass leaves. An example is *Pediocactus papyracanthus* of the southwestern United States which grows among and resembles the dead leaves of the blue gamma grass *Bouteloua gracilis*. In southern and eastern Africa several species of succulents grow among dead shrubs. They are greyish in colour, often wrinkled and twisted, and bear a close resemblance to dead branches. Matching dead grass and branches is no doubt of survival value to edible plants growing in places where browsers and grazers are common but where living vegetation is seasonally scarce; most grazing and browsing animals eat only living leaves and stems.

In Australia, and to some extent in Africa and North and Central America, various species of mistletoe bear a close resemblance to their host plants. The leaves of the mistletoes are the same shape as those of their hosts and there is a general similarity in growth form. The advantage of the similarity between parasite and host may be that the host contains chemical compounds in its tissues which protect it from plant-feeding animals. By seeming to be part of the tree, the mistletoe is ignored by plant-feeders.

Camouflage in the green and brown world

The very existence of camouflaged grasshoppers, mantids, pupae, caterpillars, and many other animals indicates intense selection by predators: if there were no selection animals would not harmonize with their background. Camouflage is a primary defence; once discovered camouflaged animals are almost invariably eaten, but as we shall see later, many animals, including the underwing moths, have a secondary means of defence which is put into operation following discovery.

Bright sunlight passing through vegetation, especially in thick woodland, results in intense illumination in some places and deep shade in others. Sun flecks illuminate parts of leaves and other objects and so the green and brown background in bright sunshine

The colour of the chameleon prawn *Hippolyte varians* is a ghostly pale blue at night. During the day it also matches its background and can change from brown and red to green. It is green when among sea lettuce.

is more patchy and varied and has greater contrast than the same background on a cloudy day with the sun obscured. Black and white photographs taken from the same position in bright sunshine and when the sky is overcast admirably illustrate this effect. For us it is more difficult to find caterpillars in bright sunshine than when the vegetation is more evenly illuminated, and predators may have the same problem. But it is moving, especially flying, prey that gain most from the patchwork of sun flecks and shade: a butterfly flying through the woods constantly passes through brightly lit and heavily shaded areas and is easily lost from view. The speckled wood butterfly *Pararge aegeria* has brown wings with pale yellow spots which look very like sun flecks. It flies and perches in those parts of a wood where sun flecks and shade occur in confusing array and immediately stops flying when the sun is obscured. And as if to emphasize its deceptive coloration, individuals that appear early in the season when the woods are less shady have bigger yellow spots than those that fly in the summer.

Some tropical butterflies have partly transparent wings and as they fly they repeatedly seem to disappear and reappear and are particularly hard to follow. Most butterflies are active in sunshine and immediately assume a resting position when the sky becomes overcast even though there may be no drop in temperature. The most likely explanation of this behaviour is that it is safer to move in sunshine.

Although butterflies may evade predators by restricting their flying to periods of bright sunshine, for most small animals movement is dangerous. It is movement that is used by all manner of predators as a means of detecting prey. Nevertheless the diversity and patchiness of green and brown vegetation, especially when subjected to varying illumination, has been and still is the main background for the evolution of camouflage. Colour, markings, counter-shading, and behaviour combine to enhance deception in the green and brown world but, as we shall see in the next chapter, there is more to it than this: there are other means of deception.

4 More on camouflage

Not all camouflaged animals match the greens and browns of leaves, twigs and bark. Extensive areas of the world support little vegetation because it is too cold or too dry; there are also sand and shingle beaches and rocky outcrops where few plants can grow. Human activities, especially mining and gravel extraction, create land temporarily devoid of vegetation. At high latitudes and altitudes winter snow transforms a drab landscape to glistening white. There are animals in these environments although they are necessarily less numerous than where there is plenty of vegetation; many of them are camouflaged.

Some insects and spiders live on and look like flowers; they are brightly-coloured but superbly camouflaged. And, even though they live in a predominantly green and brown environment, a few species, chiefly insects, resemble inedible objects like bird droppings; they are conspicuous but tend to be overlooked by predators because they do not look like food.

There are some other, rather more bizarre, forms of camouflage which will be mentioned in this chapter as well. One is pretending to be dead: a risky but often effective means of deception. Another is the camouflage some animals achieve by decorating themselves with debris, and a third is the resemblance of the eggs of parasitic birds, such as cuckoos, to those of their hosts.

Matching bare ground

Daytime temperatures in deserts are often high and, in the absence of a substantial cover of vegetation, many small animals hide beneath stones or in holes in the ground. At night when it is cool, even cold, these animals emerge into the open and go about their activities. Nevertheless many nocturnal as well as diurnal desert animals, including foxes, mice, bats, lizards, snakes and grasshoppers, are paler in colour than their counterparts in the green and brown world of wetter regions. The colours and patterns seem to blend with the desert background. But why are desert bats which roost by day in dark caves paler than those that live in forests, and why are burrowing animals like moles also paler? There is no obvious answer; one suggestion is that the pale colours are the result of loss of pigment in hot, dry air, but this is unlikely: desert mice bred for many generations in laboratories keep their characteristic colour. Countershading is well-developed in desert animals. Pale, often white, underparts are a feature of an array of species, including mammals, birds, snakes, lizards, spiders, woodlice and centipedes; again the list includes both nocturnal and diurnal animals and does not help us in deciding if we are dealing with camouflage. But mice of the same species that occur on desert soils of different colours faithfully match

The horned lizard *Phrynosoma cornuta* of New Mexico matches the desert in which it lives.

the local background, suggesting that they are indeed camouflaged.

There are also desert animals which are conspicuously black or black and white and which obviously do not match the background. Black in particular occurs in an odd assortment of birds, including larks and chats, in many beetles and some flies. Several interpretations for this unexpected coloration exist. One is that the colours are in some way associated with temperature regulation. Another interpretation is that the animals have warning coloration. It is also possible that, because in deserts individuals of a species tend to be widely dispersed, there is a need for conspicuous coloration to help in locating other individuals for mating. The significance of coloration in desert

animals is a matter of controversy. Camouflage may be important in some species but for others temperature regulation and finding mates may be at a premium. Whatever the explanation the virtual absence of green as a colour in desert animals is striking and confirms the importance of green camouflage in leafy environments.

Camouflage is more convincing in many of the animals that occur on gravel, sand, pebbles and bare earth in non-desert regions. In Europe the little ringed plover *Charadrius dubius* nests on banks and islands in newly-created gravel pits. The nest is nothing more than a shallow depression and the four eggs are pale buff, speckled with brown. They are extremely difficult to see against the background of pebbles and stones. The bird itself has irregular bands of brown, black and white across its head, neck and back. When sitting on its eggs among large stones of varying colours, it is hard to discern, for the bands of contrasting colours break up its outline most effectively. Similar disruptive coloration is seen on the killdeer *Charadrius vociferus* an American bird which nests on pebbles. Many other birds, including larks and waders, nest on bare ground and have similarly camouflaged eggs. There are also spiders, grasshoppers, and beetles that look like stones. These are generally buff-coloured, and very uniform in coloration, the legs, eyes and antennae all being the same shade as the rest of the body, and all evenly speckled with darker brown. The shape of the animal is often modified to give a square or rounded outline when at rest.

The European grayling butterfly *Hipparchia semele* occurs on heathland and chalk-downs in places where there is bare ground. The butterfly has two black-ringed, white-centred, eye-like spots near the outer edge of the underside of each forewing and, on alighting on the ground, displays one or both of these spots for a few seconds. Most butterflies can escape from predators which strike at their wings. By first displaying the eye-spots a grayling gives a waiting predator, a lizard perhaps, a chance to attack it in an unimportant part. If attacked it escapes by flying away with a slightly damaged wing. If it is not attacked, the grayling covers its forewings with its camouflaged hindwings, turns round to orient its wings towards the sun, and then leans to one side slightly, so eliminating its own shadow. (This butterfly inhabits the northern hemisphere, where the sun is never directly overhead, hence it must lean to one side to avoid casting any shadow.) It then resembles the ground and is difficult to see provided it keeps still.

Succulent plants resembling stones and pebbles are common in the arid

A grayling butterfly alights on bare ground and first exposes eye-spots on the underside of the forewings. This is an invitation for a predator to attack a non-vital part of the body. After a short time, it lowers its forewings and the eye-spot disappears. Finally it leans over to one side and so eliminates its own shadow.

regions of southern and eastern Africa and there are also examples in several species of North American cactus. In Africa, stone and pebble camouflage is known from six different plant families but is best developed in the Mesembryanthemaceae in which 23 genera and over a thousand species are involved. Species in the genus *Dinteranthus* have thick leaves and an underground stem; the leaves grow only a few centimetres above the ground and, depending on the species, are greyish-green or reddish-brown. They blend well with their stony surroundings. Interestingly all the plants that look like stones have conspicuous animal-pollinated flowers.

The flowers are short-lived and appear mainly in the wet season when vegetation is more plentiful. The resemblance to stones results in the plants being overlooked by browsing animals. There are also succulents which resemble the bare soil on which they grow. This is achieved by the possession of sticky, glandular hairs on the upper surface which accumulate particles of dust to an extent that the plant is hidden and matches the bare soil. Most of these species also belong to the Mesembryanthemaceae and occur in southern Africa, but at least three species of cactus in the deserts of North America also accumulate surface dust.

Matching the snow

A snow-covered landscape is not a good environment for active invertebrates such as insects and so they hibernate out of view in secluded places. But warm-blooded mammals and birds remain active and many of them develop white, or partially white, fur or feathers in winter. As with the green and brown butterfly pupae, we can ask 'how' and 'why' questions about the seasonal development of white fur or feathers.

The ptarmigan *Lagopus mutus* lives on high ground in Europe and North America in areas which normally receive a substantial winter snow cover. In all seasons the wings and belly of the ptarmigan are white; in the breeding season the upperparts of the male are mottled black and brown while those of the female are tawny; in autumn both sexes are greyer. In winter both sexes are pure white with the exception of the black tail which is largely hidden from view by white tail coverts (feathers which cover the tail when at rest). In this plumage the birds blend well with the snow. Similar seasonal changes in plumage occur in several species of grouse living at high latitudes in Eurasia

The eggs of the little ringed plover blend with pebbles and stones while the bold black and white head pattern breaks up the outline of the bird and is an example of disruptive coloration.

Willow grouse in winter (above) and in summer (opposite).

and North America. In the willow grouse *Lagopus lagopus* laboratory experiments show that it is the length of the day which affects the activity of hormones and produces the change in colour: by keeping birds in artificially controlled illumination, dark or white plumage can be produced out of season. Birds exposed to about 16 hours or light, simulating the long days of summer, moult their white feathers and adopt the brown summer coloration; those exposed to less than 12 hours light, as in the short days of autumn and early winter, moult into a white plumage. In this species temperature seems to play no part in initiating colour changes.

Several species of hares living at high latitudes or high altitudes in Europe and North America are brown in summer and white, or partially white, in winter; this is also true of many stoats and weasels. In the species *Mustela erminea*, known in Britain as a stoat but in America as a short-tailed weasel or ermine, individuals living in the north

or high on mountains are more likely to turn partially or completely white than those living in areas with a milder climate. Experiments show that moulting of the fur is controlled by daylength acting on hormones. Temperature, again, does not seem to affect the moult from one colour to another. But the changes occur at high altitudes as well as northerly latitudes, and an animal from high ground, if taken to low ground at the same latitude, still undergoes a moult from dark to white with the onset of winter. This suggests that the capacity to change colour is under genetic control, and that the highland animals are genetically different from the lowland animals in this respect.

Thus in birds and mammals that change colour in winter the 'how' question is answered and, although it is tempting to answer the 'why' question in terms of camouflage against a snowy background, there is still the possibility that white fur or feathers are better for conserving heat than dark ones.

Pebble plants like this *Dinteranthus puberulus* grow in arid stony desert. They are overlooked by browsing animals.

Looking like bird droppings

The Chinese character *Cilix glaucata* is a small European moth which by day sits on the upperside of leaves. The forewings are white with a grey-brown oval patch in the middle, and when the moth is at rest they obscure the hindwings. The patch is marked in places with a suggestion of silver or bluish-silver. This is not an easy colour pattern to describe; it is much easier to say that when sitting on a leaf the moth bears a striking resemblance to a bird dropping. Although conspicuous it does not look like anything edible: a bird would not normally examine its own droppings as a possible source of food. Resemblance to a bird dropping is the moth's primary defence against predators; if it is discovered and inspected it promptly falls to the ground and is lost in vegetation; this is its second defensive strategy.

The young caterpillars of several species of swallowtail butterflies also resemble bird droppings. By day they sit motionless on the upperside of leaves and although conspicuous are easily overlooked — unless of course you happen to be searching for bird droppings. In one West African species, *Papilio menestheus*, the young caterpillar rests in a bent position which further enhances its resemblance to a bird dropping. If, however, a swallow-

Opposite A young caterpillar of a West African swallowtail, *Papilio menestheus*, sits in full view on the upperside of a leaf and both its coloration and its bent posture enhance its resemblance to an inedible bird dropping.

tail caterpillar is discovered, a novel means of secondary defence is put into operation. The caterpillars possess a special gland, the osmeterium, located just behind the head. It consists of two horns formed by folds of the neck membrane and is normally hidden beneath the cuticle. If a caterpillar is prodded, perhaps by an inquisitive bird, the gland is suddenly pushed out and a strong odour is produced from the two horns which glisten with secretion. The secretion is known to deter ants which are potential predators of young caterpillars but whether it is powerful enough to deter birds and other larger predators is not known. Thus instead of falling to the ground as a means of secondary defence — which would be hazardous as the caterpillar might be unable to get back to its food-plant — it stays put and produces a defensive secretion. Fully-grown swallowtail caterpillars are green with dark and pale markings and resemble the leaves on which they feed. They are too big to retain convincingly the resemblance to a bird dropping, but they still push out the osmeterium and produce the same strong-smelling secretion whenever provoked.

Young caterpillars of an African moth, *Trilocha kolga*, are black and white and gregarious. They rest on the upperside of leaves of the food-plant and where several occur together it looks as if birds have been roosting in the foliage above and defecating on the leaves below. When fully-grown they are brown and still rest on the upperside of leaves but become solitary; they then resemble a dropping of a large bird or perhaps a lizard. Looking like an inedible bird dropping has advantages provided there are not too many species that evolve the same strategy. Some spiders appear to have this kind of camouflage, wrapping their legs tightly around their bodies to give a rounded shape, and the pupae of some moths and butterflies also look like bird droppings.

Looking like flowers

In most places flowers are either widely dispersed or temporarily and seasonally abundant. In tropical forests in particular, flowers are scattered here and there among lush greenery; only occasionally is there a tree bearing a profusion of blossom. In Europe, bluebells carpet the woods for a brief period in spring but soon die back and become overgrown with summer greenery. In North America, the yellow of goldenrod flowers is conspicuous in grassland in autumn. Even a desert can be transformed into a blaze of colour soon after heavy rainfall. There are many examples of short-term colour dominance by flowers but most, even the large and brightly coloured ones, are relatively inconspicuous among the abundance of green leaves.

Nevertheless, certain insects, spiders and other invertebrates live on flowers and are camouflaged to match the colour of the flowers. The caterpillars of a variety of species of butterflies and moths are chewers of petals and other flower parts. In Britain the ling pug moth *Eupithecia gooseniata* occurs on heaths and moors where its caterpillars feed in late summer on the flowers of heather. The caterpillar is short, stumpy and pinkish, with dark marks along the back and yellowish lines marked with brown along the sides; the head is dark olive and marked with white. These characteristics combine to camouflage the caterpillar among the flowers of heather. The darker markings act as disruptive coloration breaking up outline and shape.

Caterpillars of the double-striped pug moth *Gymnoscelis rufifasciata* are yellowish-olive, reddish-olive, or rusty red, with a yellow stripe along the side and, along the back, a dark line and a row of dark spots. They feed on the flowers of hawthorn, holly, gorse and broom; also on the purple flowers of buddleia *Buddleia davidii*, a Chinese plant introduced to Britain in 1896. The colour and pattern of these caterpillars make them well camouflaged among the greenish flowers of holly and hawthorn or the yellow flowers of gorse and broom. The variations in colour of double-striped pug caterpillars are associated to some extent with the colours of the flowers on which they feed. Those on buddleia are slightly purplish, which may be a recent adjustment to feeding on the purple flowers of a new food-plant.

A North American moth, *Schinia florida*, rests by day on the flowers of the evening primrose *Oenothera biennis* which is also the food-plant of its caterpillars. The moth is conspicuously pink and yellow. When resting in a head-down position the yellow tips of the forewings resemble the petals of evening primrose while the pink sections of the forewings blend with the colour of dying flowers below. If given a choice the moths prefer evening primrose to other yellow flowers. Even if the flowers are obscured with a muslin bag the moths still select the evening primrose in preference to other yellow flowers. Evidently *Schinia florida* responds to the characteristic scent of the evening primrose and by doing so almost invariably selects the correct background against which it can blend.

A similar North American moth, *Schinia masoni*, rests on the flowerheads of the daisy-like plant *Gaillardia aristata*. The flowerheads have yellow centres, each surrounded by a red disc and then by an outer circle of yellow 'petals' (strictly speaking they are ray-florets). The moth's forewings are red and its head and thorax yellow. Typically it rests with the red wings matched against the red disc and the yellow head and thorax against either the yellow centre or the outer yellow florets. The red disc produces nectar and to gain access to it with its tongue the moth has to rest in the 'correct' camouflaged position. In this species a necessary feeding position has led to the evolution of distinctive red and yellow concealing coloration.

Small, crab-like spiders of the family Thomisidae occur in many parts of the world. All are predators of insects which they catch by waiting motionless on vegetation. Some species wait on flowers to catch nectar-feeding bees, butterflies and hoverflies. Their colours and patterns blend perfectly with the flower colours and patterns. When a flower dies the spider moves to another; if it is of a different colour the spider sometimes changes colour and harmonizes with the new background; exactly how it does this is not known. One species of crab spider, *Thomisus onustus*, varies in colour between pink and yellow. It occurs on heathland and pink individuals match well the pink heather flowers on which they wait for nectar-feeding insects. When an insect approaches a flower the spider positions itself so that it can seize the insect's head, an operation it performs with speed and agility. Grasping the insect in its mouthparts, the spider then sucks liquid from its prey. The dead prey appears undamaged and remains on the flower after the spider has finished feeding. Butterfly-collectors, cautiously approaching a flower with a prize specimen perched on it, are sometimes astonished because the butterfly does not fly away; close inspection reveals that it is dead and a careful search usually leads to the discovery of *Thomisus onustus* or another of the flower-dwelling species of crab spider.

Camouflage in crab spiders could function either as a means of concealment from their prey or as a means of deceiving their own predators, or both. An enterprising spider-watcher once devised an experiment to test which of these possibilities is the more important. He arranged sixteen dandelion flowerheads at equal distances from each other on a square of lawn. In the centre of every second dandelion he placed a black pebble about the same size as a crab spider, and in the remaining eight dandelions a yellow pebble especially selected to match the colour of the flowerhead. In the next half an hour he recorded the visits of nectar-feeding

insects to the dandelions and obtained the following results: 10 honeybees, 42 hoverflies and 4 other insects to the flowerheads with yellow pebbles; 1 honeybee, 4 hoverflies and 2 other insects to the flowerheads with black pebbles. This gives 56 visitors to the dandelions with yellow pebbles and only 7 to those with black pebbles. Besides those that actually alighted, honeybees and hoverflies were seen to inspect repeatedly and then withdraw from flowerheads with black pebbles but settled without hesitation on those with yellow pebbles. This simple but ingenious experiment shows that nectar-feeders are put off by something they can see; it also suggests that camouflage in crab spiders is more likely to be an adaptation to conceal predator from prey than prey from predator. It does not, of course, totally rule out the possibility that a crab spider's coloration also affords it a degree of protection from predators.

In the tropics there are species of mantids, sometimes known as flower mantids, which, like crab spiders, lurk on flowers and seize insects that visit them for nectar. Some of these mantids are brightly-coloured but, although large, are difficult to see unless they move. As well as matching the colour of the flowers, often pink or white, the segments of the legs are broad, flattened and curved in outline to give petal-like shapes. It seems likely that their coloration conceals them as they wait for prey to arrive.

Feigning death

When at rest the European pale prominent moth *Pterostoma palpina* looks exactly like a slightly weathered chip of wood. If prodded the moth falls over but does not move its wings, legs or antennae; it still looks like a lifeless chip of wood. Birds encountering an apparently immobile and lifeless object such as this would probably leave it alone as it bears little resemblance to anything remotely edible.

Some tropical mantids rest on and look like flowers. This may protect them from predators or conceal them from prey that approach to feed on the 'flower'. This mantid, *Hymenopus coronatus,* from Borneo is pretending to be a flower.

Active animals, including certain beetles, grasshoppers, mantids, spiders and snakes, may become temporarily immobile if attacked. The animals most famous for feigning death when attacked are the North American opossums *Didel-* *phis* whose behaviour often results in a would-be predator going away and trying elsewhere. Some predators, including cats and mantids, attack living animals only; they must kill, and will not touch a food item unless they have

Above African flower mantid, *Pseudocreobotra wahlbergi,* both as an immature (left) and an adult (right). The adult has caught a butterfly that came to the flower for nectar.

Left To gain access to nectar, this moth, *Schinia masoni,* has to sit in the 'correct' position. A necessary feeding position has led to the development of distinctive red and yellow coloration which matches the flower, *Gaillardia aristata.*

Right An unsuspecting bee approached a flower for pollen and nectar and was seized by a crab spider which had been hidden in the flower.

killed it themselves. Feigning death is a special form of camouflage, but it is risky because not all predators are put off by prey which seems to be dead.

Camouflage by decoration

Aphids excrete copious quantities of sugary honeydew which is consumed by many organisms, among them ants. The ants do not attack the aphids but 'milk' them for their honeydew and guard them from predators. To prey upon aphids a predator must come to terms with the attendant ants, which are ferocious insects, often equipped with an unpleasant bite. This is done in a variety of ways, including the use of disguise. The larvae of the North American lacewing fly, *Chrysopa slossonae*, live among and feed on the woolly alder aphid *Prociphilus tessalatus* and cover their bodies with the waxy 'wool' which they remove from the aphids. Lacewing larvae are then mistaken by the ants for aphids and left alone. But if a human investigator strips the lacewing larva of its wool it is immediately attacked by the ants and removed from the aphid colony.

In Africa it is common to see what appears to be a small and densely packed cluster of dead ants moving rapidly on a wall or termite mound. Close inspection reveals that the ants are packed on the back of a large bug. The bug, *Acanthaspis petax*, preys upon ants and, after sucking fluids from their bodies, decorates itself with corpses. The bugs also decorate themselves with particles of soil and small stones, perhaps when ants are not readily available. Similar species occur in many other tropical regions.

Certain caterpillars of geometrid moths cover themselves with lichens, and some marine animals, particularly gastropod molluscs and crabs, also decorate themselves. Many spider crabs, (not to be confused with crab spiders) decorate their carapaces with living sponges, algae and sea anemones, and hence do not look like crabs. These

Nothing looks less like a bug than a small heap of ants. By decorating itself this Philippine bug evades predators which are unlikely to investigate such an unpromising food source.

organisms take a hold on the carapace and continue to grow there. But perhaps more impressive are the so-called carrier shells of the family Xenophoridae. These are marine gastropod molluscs which decorate their spiral shells with debris including other, smaller shells. One species, *Xenophora conchyliophora*, which lives in shallow water off Florida, has been observed decorating itself when kept in an aquarium. It uses debris from the sea bottom, particularly the empty shells of bivalves, sticking them in an upside-down position on top of its own shell. It pushes the items into place with its head and proboscis, and they are then cemented to the shell with a substance produced by the snail's body. As the shell is decorated gaps are filled in with smaller items of debris until a complete covering is achieved. After attaching a large item to itself *Xenophora conchyliophora* remains still for up to ten hours, presumably to enable a firm bond to be formed. When fully decorated the mollusc matches exactly the sea bed, the upside down bivalve shells being a particularly effective disguise. Other species decorate themselves with pebbles.

Cuckoo eggs

Most species of cuckoos lay their eggs in the nests of other birds and the young are raised by unwitting foster parents; brood parasitism of this type has been independently evolved in several other groups of birds including the New World cowbirds, the African and Asiatic honeyguides and some species of African weavers. The parasitic association involves deception of the host to an extent that it is fooled into raising offspring which are not its own; the parasitic bird has no parental responsibilities. In many of these brood parasites the eggs match in coloration and size the eggs of the host. Some cuckoos, including the Eurasian cuckoo *Cuculus canorus*, utilize many different species of hosts each of which has distinctively coloured and patterned eggs. Individual female cuckoos parasitize the same species over and over again, laying up to twenty eggs in a season, normally one in each nest. Each female produces eggs that match in coloration those of its normal host, although in some places, including Britain, where many different hosts are used, the match is not especially accurate. If an egg is laid in the 'wrong' nest and is a poor match, the nest is often deserted by the potential host or the cuckoo's egg is thrown out. Thus the hosts constantly exert selection which eliminates those genotypes that produce poorly matching eggs. Some hosts are more discriminating than others.

A study was made of the eggs of *Cuculus canorus* found in the nests of a wide variety of hosts in the hills or northeast India. The eggs could be divided into two categories: those in the nests of normal hosts, and those in the nests of abnormal hosts. Most of the eggs in the nests of normal hosts matched the coloration of the hosts' eggs while many of those in the nests of abnormal hosts were a poor match. Of 1642 normal hosts' nests, 135 (8 per cent) deserted, while of 298 abnormal hosts' nests, 72 (24 per cent) deserted. This result is a splendid demonstration of natural selection: poorly matching eggs in the wrong nests are much more likely to perish than well-matching eggs in the right nests.

The cuckoo is a very strange bird. The nestling is brought up by a totally different species and has no contact with its own parents. Adult European cuckoos leave for winter quarters in Africa ahead of the fledged young which migrate south later and may or may not encounter older cuckoos while in Africa. The birds return to Europe in the spring, they mate, and the females seek out and lay eggs in other birds' nests, usually picking hosts with eggs of the same colour and pattern as their own. How do they choose the right host? And, just as importantly, how do they manage to lay eggs of the same coloration again and again? Female cuckoos always produce the same colour egg, no matter what male they mate with.

In animals there is one pair of chromosomes that is responsible for sex determination. The pair is either made up of two identical chromosomes (and therefore called XX) or two different chromosomes (called XY). In birds the XY complement occurs in females and the XX in males. If a male cuckoo mates with a female all the resulting eggs laid by the female are the same colour, no matter what the genetic make-up of the male. More importantly, all the cuckoo's daughters will lay eggs of the same colour. This means that the genes which determine egg colour are likely to be located on the Y chromosome. Because the X chromosome does not have any genes which match those on the Y chromosome (as is the case with other chromosome pairs) the Y chromosome's effects are always expressed. Such Y linkage, as it is called, is rare in animals, but it does mean that a female can transmit a fixed trait to her female offspring no matter what sort of mate she uses. Unfortunately, this must remain a hypothesis because, as it will be appreciated, the chances of breeding cuckoos in captivity and working out

Above The grotesque expression of a grass snake feigning death when confronted by a predator.

Right A spider crab can look like the sea bed by attaching sponges and sea anemones to its carapace.

A cuckoo's egg in the nest of the great reed warbler. It is a good match.

the genetics of egg coloration are remote: their behaviour is far too complex for that. Assuming, however, that the hypothesis is correct and that the female cuckoos have a genetic system that fixes individual egg coloration, we are still left with the question of choosing the correct host.

Birds such as geese are well known for their ability, when young, to become imprinted on animals that are not geese. A gosling will attach itself to the first animal it sees after hatching and regard that animal, whether goose, dog or human being, as its mother. There is no reason why cuckoos should not have a similar ability to become imprinted. During the brief period that they have

contact with their foster parents, while being fed in the nest and for a short while after fledging, young cuckoos could become imprinted so that they remember their foster parents and in the next year seek out and lay eggs in the nests of the same species. Once this has been done in the first breeding season the female cuckoos could become further imprinted and do the same in subsequent seasons.

5 Being different

The North Atlantic brittlestar, *Ophiopholis aculeata*, is extraordinarily variable in colour and pattern. It may be red, purple, black, yellow, orange or pink, and both the body-disc and the arms are affected. The disc may be with or without a conspicuous border; there may be spots in the area where the arms join it, a single central spot, or spots arranged in a variety of ways around the edge; there may be a large five-pointed star-shape, or various types of speckling. Every combination and permutation seems to exist and it may be necessary to examine a hundred or more individuals before two are found that are alike. The shell of the butterfly clam *Donax faba* a bivalve mollusc on sandy beaches in East Africa, may be white, pink, purple, mauve, red, grey or brown; some are banded with darker colours, others are unbanded. It spends most of the time buried in the sand and is exposed only when waves break on the tide-line. Both the brittlestar and the clam are very abundant where they occur. Neither are camouflaged against the background in which they live, except when very young when most of the colour forms of the butterfly clam resemble their background of sand grains.

The flat periwinkle *Littorina obtusata* which lives among seaweed on the North Atlantic coasts, is another mollusc which has many colour forms of the shell. It varies through greens, black, greys, browns, reds, yellows and white, and may or may not be banded. The frequency of the different colour forms varies from place to place and some populations are more variable than others. In the flat periwinkle, as in certain species of moths, bugs and land snails with similar variation, there are colour forms that appear to match the background and forms that contrast with it. The green and brown forms have an uncanny resemblance to the bladders of the seaweeds bladder wrack *Fucus vesiculosus* and knotted wrack *Ascophyllum nodosum* on which they are found. Other colour forms of the flat periwinkle, such as red and yellow, are quite conspicuous. Can the spectacular variation in such animals as the brittlestar, butterfly clam, and flat periwinkle be explained in terms of natural selection by predators?

Consider a bird hunting for periwinkles on a seaweed-covered rocky shore. Because of previous experience it forms an image of what it is looking for and goes on searching for the same items, poking among seaweed and trying anything that resembles suitable food. As it feeds it takes proportionately more of the periwinkles with the shell colour that it has already eaten, often ignoring or overlooking those of a contrasting colour. This means that periwinkles with a contrasting shell colour are at an advantage provided, of course, they are not too common; if there were too many they would soon become part of the bird's search-image. Now imagine a great many birds feeding in the same way over a long period of time. Each feeds selectively, search-images formed by different individuals change from day to day, and different coloured periwinkles are eaten or missed, depending on the previous experience of the predators. Some shell colours of the periwinkles match the colours of the seaweed, others may not; what is important is that there are a lot of colours and that they contrast with one another. The chance of survival of each form depends on its frequency in the population; in other words there are advantages to being relatively rare and disadvantages to being too common. In time a state of balance is achieved, the frequency of each form being determined by the combined effects of predators. This explanation is entirely theoretical and is based on what is believed to be the way in which some predators search and its selective effect on the coloration of prey. Is there evidence to support the theory?

In the 1940s an experiment was devised in Britain to investigate whether the rudd *Scardinius erythrophthalmus*, a predatory freshwater fish, feeds selectively on the water-boatman, *Sigara distincta*, a small aquatic bug. The bugs vary from light to dark brown. The predators and prey were kept in an aquarium with brown sand on the bottom. Bugs that contrasted with the sand were taken more frequently than those that matched it, a result that shows the effectiveness of camouflage. But more interestingly, the number of each form taken by the rudd varied with their frequency: no matter what shade of brown they were, those at low frequency were taken less often than those at high frequency. In other words the experiment showed that the chance of a bug of a particular shade of brown being eaten by a rudd was dependent on how rare it was relative to other bugs

of different shades of brown. This is an example of what is known as frequency-dependent selection. Such selection could be expected to occur whenever predators form search images as a result of previous experience in finding prey. In this experiment the prey varied in tone rather than in colour and pattern and it might be that the contrasting colour forms found in brittlestars, butterfly clams and periwinkles are even more advantageous.

A rather simple experiment which has often been conducted is to offer wild birds artificial 'prey' of various colours. Pellets of pastry are coloured with edible dye and placed where birds come to feed. If for a period of days the birds are offered pellets of only one colour and then later they are offered a choice of two colours in equal quantities, they will take more pellets of the original colour. This occurs regardless of the colour of the background. If the pellets are offered at different frequencies the birds take proportionately more of the commonest colour. Such an experiment can be tried on wild birds that come to feed on the lawn. The results are not always conclusive, especially if there is a large flock of birds present but if, for example, there are just a few starlings, a clear-cut result can sometimes be obtained. Admittedly the circumstances are somewhat abnormal but, with the difficulty of obtaining direct evidence of this sort in truly natural situations, the results certainly suggest that birds form search images and that frequency-dependent selection is a reality, even in the absence of camouflage.

Around Kampala in Uganda the land snail, *Limicolaria martensiana*, occurs in discrete and isolated populations in gardens and plantations of gum trees. In this area there are four distinct colour forms of the shell: 1) entirely pale buff except for short streaks at the edge of the whorls, 2) pale buff with the streaks fully developed but very faint, 3) entirely pale buff without markings, and 4) a buff background, heavily and intricately

morphism, just the streaked form. Where the populations are most dense there are more diverse colour forms, but where the snail is scarce the heavily streaked form predominates and the other forms tend to disappear.

The streaked form of *Limicolaria martensiana* looks as though it matches the background for, in the habitats where the snails live, bright sunshine produces contrasting patches of light and shade into which the snails blend. But the buff forms are conspicuous and look very different from those that are streaked. Many predators eat the snails, among them the open-billed stork *Anastomus lamelligerus*. This bird is a specialist snail-eater, but whether it forms search-images and takes the colour forms selectively is not known. All the same, the chances are that the variations in colour-form diversity are maintained by frequency-dependent selection. In places where the snails are common there is a greater chance of being discovered and eaten than where they are rare, and so the variation in colour tends to be greater in these places. Evidence in support of this suggestion

Above Polymorphic periwinkles on bladder wrack. Some may be camouflaged, others may survive simply because they contrast with the others and are therefore overlooked by predators.

streaked with black. There are thus three more or less pale buff forms and a heavily streaked form which looks very different from them. In some places the snails are extremely abundant and occur at high densities: more than 100 per square metre. Here the streaked form comprises between 40 per cent and 60 per cent and the three buff forms make up the remainder. In other places the snails occur at low densities, only about 20 per square metre and, although all four forms are present, the streaked form comprises about 75 per cent. At one place the snails occur at a very low density of only 5 per square metre and here the streaked form comprises 93 per cent of the total and only one of the buff forms is present. Scattered throughout the area are places where the snails are scarce, less than 1 per square metre, and in these places there is no poly-

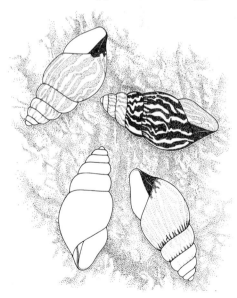

The forms of the African land snail, *Limicolaria martensiana*. The heavily streaked form is probably camouflaged while the three paler forms contrast with it and, by being different, are often overlooked by predators.

Left Three kinds of brittlestar. By being different, predators are less able to recognize them.

came when shrubs and tall grasses were cleared from one place where the snails live, and the habitat became more open and dry. The population density fell from 26 per square metre in 1963 to 2 per square metre in 1966, and at the same time the frequency of the streaked form rose from 68 per cent to 79 per cent.

Experimental evidence which shows that if the population density of prey animals is high they are more likely to be found by a predator comes from an investigation with carrion crows *Corvus corone*. Chicken's eggs were spread on the ground at intervals of either 50 or 800 centimetres. Carrion crows, which are notorious egg predators, found 89 per cent of the eggs when these were spaced at intervals of 50 centimetres but only 19 per cent when they were spaced at intervals of 800 centimetres. That carrion crows also form search-images of their prey was shown by another experiment in which painted mussel shells with pieces of meat placed beneath them were spread over the ground; the crows found more shells if they were all the same colour than if they were of three contrasting colours.

Hence diversity in colour and pattern can be regarded as a defence tactic against predators that learn by experience and form search-images. This is the simplest explanation for the extra-ordinary variation in the brittlestar, the flat periwinkle and other small, polymorphic animals, that are apparently not camouflaged, but it may not be the only one.

One of the features of scientific research is that once a problem has been experimentally investigated new questions emerge. Sometimes it seems that the more we know the more difficult it is to be sure that we have the right answer and that the less we know the easier it is to offer an explanation. Although scientists tend to favour simple unifying solutions to problems, alternative hypotheses arising from new questions are welcomed. The approach is to investigate and then reject each hypothesis until a single solution is found. But, although a single solution may be a scientist's goal, it does not always turn out like this. Indeed, the outcome of intensive investigations over the past thirty years on the significance of polymorphism in colour and pattern in the land snails, *Cepaea nemoralis* and *Cepaea hortensis*, has left us with at least eight different explanations of the phenomenon. Before the research was started there had been only one explanation which, in the event, was wrong.

Both snails are common in Europe. They tend to form discrete colonies in

Almost everywhere they occur, the land snails *Cepaea hortensis* (left) and *Cepaea nemoralis* (right) exist in a puzzling array of colour forms.

woodland and open country, and often both species are present in a single colony. The shell of *Cepaea nemoralis* is usually some shade of yellow, pink or brown and may have from one to five dark brown bands or may be unbanded. Adjacent bands may be fused in a variety of ways and sometimes all five are joined together. *Cepaea hortensis* is similarly polymorphic but is less variable than *Cepaea nemoralis*. The frequency of the colour forms in both species varies markedly from place to place: even over a distance of a few metres there may be striking differences.

In the 1940s, before any research was started, polymorphism in *Cepaea* was assumed to be of no adaptive significance. It was claimed to be the result of random mutation and the absence of natural selection and of no survival value to the snails. Then in the 1950s it was found that the song thrush *Turdus philomelos* takes different coloured snails at certain seasons of the year. The thrush brings snails it has found among vegetation to stones where it skilfully cracks them open, eats the body, and leaves the broken shells scattered around. By comparing the frequency of colour forms in the shells found around a thrush's anvil and living snails in the same colony it is possible to show that the thrushes are selective. Pink and brown snails are more frequent in woods and match the background better than yellow snails which are more frequent in open country. Hence camouflage against the background was put forward as the explanation of the polymorphism and the song thrush was identified as the predator. Random mutation was rejected as an explanation.

Captive song thrushes can be trained to form search-images of a particular shell colour so that, when allowed to feed on snails in a polymorphic population, they tend to select the colour form which they have already learnt to recognize. It was therefore suggested that, at least in some places, the significance of the polymorphism lies in the difference between contrasting colour forms and that although camouflage against the background may be important, it is not the only possible explanation.

Unfortunately for these two theories, polymorphic *Cepaea hortensis* occur in Iceland where there are no song thrushes. Research in Iceland and elsewhere in Europe has established a correlation between temperature and the occurrence and frequency of certain of the colour forms in both of the species. Moreover if temperature probes are inserted into living snails it can be shown that banded individuals reach a higher temperature than unbanded, and brown individuals a higher temperature than yellow. In all environments temperatures vary and another explanation of the polymorphism is that it represents an adjustment to fluctuating temperature. Given that there is a polymorphism, the occurrence of certain types in one locality and different ones at another, could be due merely to chance. If a single fertilized snail makes its way into the new and suitable area, a population may eventually be established from that one snail. The array of colour forms in the new population depends on the genes carried by the founder. Thus chance may play a part in determining what forms occur in a given locality, but the frequency of these forms, relative to one another, will probably be affected by natural selection of some kind.

Other explanations for the polymorphism have been put forward, and all have something to be said for them, but not all operate in every population or colony, and not one of them is likely to be entirely responsible at a given place or time. In *Cepaea*, being different as a means of escaping predators may be the main explanation of the polymorphism in some colonies, but may be totally unimportant in others. Exactly the same may be true in the brittlestars, butterfly clams and African land snails: we simply do not know.

6

Deception by invitation

When a grayling butterfly alights on the ground and momentarily exposes the eye-spots on its forewings it is, in effect, inviting a waiting predator to attack a non-vital part of its wings. A variety of animals, including birds, lizards, snakes, bugs, spiders, butterflies and moths have evolved colours, patterns, structures, and behaviour that divert a predator's attention and which very often result in them escaping with only minor damage or discomfort. Deception by invitation is a secondary defence strategy, put into operation once a prey has been discovered by a predator or, as in the grayling, when discovery is a distinct possibility.

Eye-spots

Many species of butterflies and moths and some mantids have eye-like markings on the wings; some turtles and caterpillars have similar markings on their bodies. These may be small black rings with white or pale yellow centres, or they may be large and elaborately coloured with some combination of black, white, red, yellow or blue. We call these markings eye-spots because they look like eyes, but we must be cautious in assuming that animals react to them in the same way. There is evidence that predators respond to eye-spots in two different ways. If the spots are large, brightly coloured, few in number, and especially if they are suddenly exposed and displayed when the animal is disturbed, a predator is startled, even frightened, and is deterred from investigating further. This means of deception is discussed in Chapter 7. (In addition some fish and birds, for example the peacock, have large eye-like spots which are important in courtship displays and social communication and have nothing to do with the deception of predators.) But if the spots are small, and especially if they are exposed in situations where there is the possibility of attack, they function as targets which misdirect the predator. Eye-spots which function in this way are located on a non-vital part of the animal, such as the edge of the wings of a butterfly. By contrast, the large eye-spots used for frightening off predators may be found on, or near, vital parts of the body.

Evidence for the effectiveness of deflecting eye-spots comes partly from experiment and partly from observation. In one experiment small eye-spots were painted on mealworms which were then offered to birds. The birds pecked more often at the mealworms with artificial eye-spots than those without them and usually aimed their pecks at the spots, showing that a bird's attention can be diverted by a target. If a large sample of butterflies is examined some of them will be found to have distinctive triangular beak-marks on the wings; these are individuals that

The eye-spot has attracted the attentions of a bird, but the butterfly has escaped unharmed.

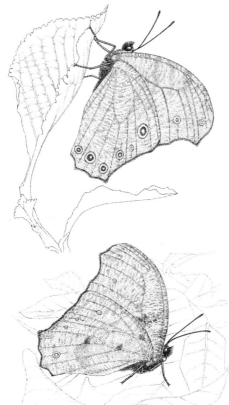

Melanitis leda, a tropical satyrid butterfly: wet season form (above), dry season form (below).

have been attacked and missed by birds. In species of butterflies with eye-spots near the edge of the wings the beak-marks are frequently near the spots and well away from the vulnerable body, an observation which also suggests that a bird's attention is diverted by a target. In another experiment artificial eye-spots were painted on butterflies' wings, the butterflies released, and as many as possible recaptured a little later on. A large number had beak marks near the spots. Absence of deflecting eye-spots might well mean that a bird would strike a butterfly closer to the body and hence kill it.

Most butterflies of the family Saty-ridae, which includes the grayling, have small eye-spots near the edge of the wing. The number and arrangement of the spots varies with the species: they may be on the upper- or underside of the wings or both, and there may be many or just a few.

The African and Asiatic satyrid, *Melanitis leda*, is unusual for a butterfly

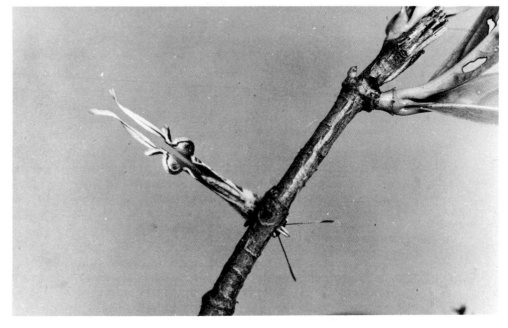

A Malayan lycaenid butterfly with a false head. The deception is particularly striking when viewed from below.

because it flies chiefly at dusk and spends the day resting among fallen leaves beneath trees and shrubs. In the forest region of West Africa, where there are clear-cut wet and dry seasons, the adult butterflies occur in all months of the year. Those that appear in the wet season have about six small eye-spots near the outer edge of the hindwings, but in dry-season individuals the eye-spots are absent, or very small and inconspicuous. The alternation of wet and dry season forms in the butterfly is as regular and predictable as the alternation of wet and dry seasons themselves. Individual butterflies do not change their coloration and so the wing pattern which develops must be programmed during the immature stages.

Once again we can ask 'how' and 'why' questions. Caterpillars reared in a dry, air-conditioned laboratory give rise to the dry season form: evidently the humidity a caterpillar experiences determines which form of the butterfly results from it. By varying the humidity intermediates between the two seasonal forms can be produced although these are rarely encountered in the wild. This experiment answers the 'how' question;

the 'why' question is more difficult and requires a certain amount of speculation.

Possibly in the dry season, resemblance to dead leaves is at a premium: the butterflies are brown and leaf-like and blend well with their surroundings. As well as the absence of eye-spots, there is much individual variation in the pattern and shade of brown on the underside; no two are exactly alike, which is in keeping with the various patterns and shades of brown of fallen leaves. In the wet season however there is little accumulation of dead leaves, and wet season butterflies look less like brown leaves. Although they are apparently camouflaged against a dark background the eye-spots are conspicuous. It might well be that because there are less dead leaves about the possession of deflecting eye-spots is a better means of escaping predators than a completely camouflaged coloration.

False heads

A blow to the head is the most effective way of killing an animal and many predators make their initial strike at the prey's head. Many butterflies in the family Lycaenidae have false heads at the tips of their hindwings well away from their true heads. The impression of a head is given by antennae-like extensions of the tip of the hindwings. These are moved up and down for a short period after the butterfly has alighted in the way that an insect's antennae often are, while the real antennae are kept still. At the base of the false antennae there are conspicuous markings, in some species eye-spots, which resemble the head itself. *Thecla togarna*, a South American hairstreak butterfly, not only has false antennae and eye-like markings but brown stripes on the whitish hindwings which converge on the false head and appear to direct attention to the 'wrong' end of the butterfly. On alighting, *Thecla togarna* immediately turns round and faces the direction from which it has arrived and so, if disturbed,

it flies off in the opposite direction to that expected. Presumably the combination of a false head at the tip of the hindwings and the unexpected flight direction confuses a predator that has watched the butterfly alight and tries to catch it. If the predator is successful the attack is probably at the false head which is disposable. Four of the five British species of hairstreak butterfly have false heads but they are not as elaborate or as convincing as those in many tropical Lycaenidae.

A Southeast Asian lantern-bug, *Ancyra annamensis*, has a most spectacular false head at the rear end. The conspicuous false antennae and eyes are large and black while the real head is tucked down and the eyes are small and inconspicuous. When disturbed the bug appears to move backwards suddenly.

Several species of snakes have brightly-coloured tips to the tail; in one species, *Calliophis maculiceps*, it is red and blue with black markings and is raised up in a curved position when the snake is disturbed. This snake is not poisonous and the false head may appear menacing to a predator or it may distract attention from the real head to the less vulnerable tail. Two heads, it seems, are better than one, at least for certain snakes, bugs and butterflies.

Distraction behaviour

Some ground-nesting birds, in particular ducks and waders, have evolved a special form of behaviour called a distraction display which they perform when disturbed from a nest by an intruder. They leave furtively but once they are a little distance away suddenly show themselves and feign injury by flapping one or both wings against the ground; at the same time they move further and further away from the nest. To the human observer this behaviour is compulsive watching; it is virtually impossible not to follow the bird and consequently lose sight of the place where the nest is located. Eventually the bird sud-

denly recovers and takes flight. If a female European nightjar is disturbed from her nest she trails her wings in a most convincing manner. Ducks often perform a distraction display on water where they are relatively safe. Injury feigning is certainly distracting to us and there seems little doubt that it is equally confusing to foxes, crows and other nest predators.

Another form of distracting behaviour, only possible under water, is shown by some cephalopod molluscs (the octopuses, cuttlefish and squid). If a cuttlefish is attacked it releases a cloud of black 'ink' into the water which diverts the predator's attention and possibly also obscures vision while the cuttlefish quietly swims out of danger. Various other marine animals produce clouds of coloured fluid when attacked which are thought to distract predators.

There are many other body forms which may be distracting to a predator. Some moths have long tails extending from the rear of the hindwings. In the African saturniid moth, *Eudaemonia brachyura*, the tails are twice as long as the moth; the tips are twisted and pale in colour, and when in flight at dusk the moth looks as if it is being followed by something else. Again this is distracting to the human observer and probably to predators as well.

Disposable body parts

Small pieces removed from a butterfly's wing by an attacking predator do not harm the butterfly. Most butterflies that have been flying for several days show signs of wing damage partly caused by brushing against vegetation and partly by encounters with predators from which they have escaped. Bits of wings are disposable and because of this many butterflies survive an attack.

Geckos, some other lizards and certain salamanders are capable of detaching their tails from their bodies

An avocet makes a desperate attempt to pretend to be injured when its nest is approached.

The American moon moth *Actias luna*. The long, twisted tails of some species of night-flying moths distract a predator's attention.

when attacked. The lizard escapes and the detached tail continues to wriggle thus diverting the predator's attention. Lizards which lose their tails in this way can eventually grow them back again.

Lures

Deception by invitation is also used by predators as a means of enticing prey: if a suitable bait is offered prey can be attracted and then captured. In the angler-fish, *Lophius piscatorius*, the first spine of the dorsal fin is very long and thin and held forward in a bent position in front of the mouth. At the tip of the spine is a lure: a double flap of fin webbing which resembles a worm. The spine itself can be moved in all directions, and functions as a fishing line.

Small fish are attracted to the lure, the huge mouth of the angler is suddenly opened, and the resulting rush of water forces the prey into the mouth which is then closed. The mouth is bordered by long, curved, backward-pointing teeth which trap the prey inside. Prey are not chewed but simply swallowed whole. *Lophius piscatorius* is a large fish of up to 2 metres long. It is flattened and spends most of the time motionless on the sea bed against which it is beautifully camouflaged. When it settles it digs itself into sand and mud and is partly hidden. The seaweed-like flaps of skin all around the edge of its body enhance its resemblance to the sea bed. Indeed, apart from the lure, which moves about in a most tempting way, the angler-fish is virtually impossible to see, despite its size. It is a truly remarkable fish: it can scarcely swim, and is almost all head and mouth with only a small body. There are angler-fish of one sort or another in seas throughout the world. Most live in relatively shallow water, but there are also deep sea species and in these the lure is luminous.

A variety of plant families contain a few species which lure and kill insects. Most grow in soils deficient in nitrogen, an essential element for the production of proteins. By digesting insects, these plants obtain protein which might otherwise be in short supply. Sundews *Drosera* are abundant in acid bogs in many parts of the world, and have leaves with sticky hairs which glisten and look as if they are covered with dew-drops or sugary nectar. An insect attracted to a sundew leaf is caught in the sticky secretion and held fast by the hairs which close over it. A special secretion is then produced which digests the insect's body so that it can be absorbed by the plant.

In Venus' fly-trap, *Dionaea muscipula*, some of the leaves are two-lobed with the midrib of the leaf functioning as a hinge. The insides of the lobes are red; an insect which is attracted and lands on them triggers off sensitive hairs which stimulate the two lobes to close. The spiny edges of the leaf lobes interlock and the trapped insect is digested. If the closing of the lobes is artificially stimulated by a tap from a pencil, the leaf soon opens again, but if an insect is present the leaves stay closed while its body is digested.

Pitcher plants, *Nepenthes* and related genera, have the whole or part of the leaf modified into a jug-like structure often with a lid attached to one side of the opening. Around the opening are markings that resemble flowers, and also a secretion of nectar. When an insect lands on a 'flower' it often slips on the edge of the jug and falls inside where it is digested in a fluid containing special enzymes. The insect is thus lured to what appears to be a nectar-produc-

ing flower and is killed. This flower-trap has been exploited by another predator. In Borneo a crab spider, *Misumenops nepenthicola*, lives inside the jug of *Nepenthes gracilis* where it makes a silken support and catches some of the insects that fall in.

A number of orchids, and some other plants, have flowers which lure insects by resembling insects, not to prey upon them but for the purposes of pollination. The flowers resemble the females of a particular insect species, generally a bee or wasp. As well as reproducing their colour and shape, the orchids with bee-like flowers have hairs on the lip of the flower which simulate those on the insect's abdomen. These flowers generally appear early in the season, before many female bees have emerged, and when the males, which emerge earlier, are unable to find females with which to mate. They are attracted to the flowers and attempt to mate with them instead. In going from one flower to another they carry pollen and achieve cross-fertilization, which is of benefit to the plants. By closely resembling a particular species of bee or wasp the orchids ensure that the pollen they get is from a plant of the same species.

Overleaf To be lured to a sundew plant can be a sticky business.

7

Intimidating the predator

Predators are wary and are easily put off by the unexpected. Birds may be excellent opportunists but they are also nervous and suspicious of anything unusual: there seems to be danger everywhere, even from items of potential prey. If a blue tit pecks at a resting moth and the moth gives an unexpected response, the blue tit is startled and the moth escapes, or at least this is what seems to happen. It is not difficult to imagine that we can recognize a frightened bird when we see one but whether we are really witnessing fear is another matter. By calling this chapter 'Intimidating the predator' the assumption is that predators can be intimidated and respond accordingly; but it must be admitted that we are on shaky ground and can only surmise what really happens when a predator seems to be intimidated.

Intimidation, or what appears to be intimidation, occurs when a predator approaches or attacks and the prey responds in a way that deters the predator or simply startles it so that escape is possible. Four different responses by the prey may be recognized: [1] sudden exposure of a bright colour pattern or a large eye-spot, [2] aggressive display in which the animal resembles something dangerous or unexpected, [3] production of a defensive secretion which has an unpleasant smell or taste, and [4] making a noise. Very often two or more of these responses are given at once. Each, in its own way, is a form of deception as the animal is pretending to be much more frightening than it really is — intimidation is mostly a matter of bluff.

Flash coloration
The European yellow underwing moth *Noctua pronuba* is common almost everywhere, especially in gardens where it spends the day lurking in herbaceous borders and vegetable patches. Its forewings vary from yellow-brown to purplish-brown and there are both paler and darker markings. When the moth is at rest the forewings cover the bright yellow-orange, black-bordered hindwings. If a resting moth is disturbed it immediately opens its wings and flies off at speed. The sudden appearance of bright colour can be mildly startling to a gardener working peacefully in his vegetable patch, and it is likely that birds searching for food in dense vegetation are similarly startled. The sudden exposure of bright colour and bold pattern in an otherwise camouflaged animal is an example of what is called flash coloration.

Underwing moths of the genus *Catocala* have camouflaged forewings and hindwings which are either brightly-coloured and boldly marked or strikingly black. These hindwings are hidden when a moth is at rest on a tree trunk. On being disturbed the hindwings are suddenly exposed and the moth takes flight. Samples of underwings of 41 different species were collected at four widely-separated places in eastern North America. These 41 species could be placed into five groups of hindwing pattern: [1] black with white bands, [2] unbanded black, [3] yellow with black bands, [4] orange-red with black bands, and [5] pink with black bands. Interestingly, although the relative frequency of the 41 species varied greatly from place to place, the frequency of the five hindwing groups remained constant. Group 1 was always rare and group 3 was by far the most frequent; groups 2, 4 and 5 occurred at intermediate frequencies but with very little variation between each of the four collecting sites. This suggests that there is some advantage in having so many of one colour and pattern, relative to those of other colours and patterns.

Why is it that, despite wide differences in the relative abundance of species, the five hindwing groups remain constant? To answer this question we must first discuss the significance of

hindwing coloration to the survival of underwing moths. Theoretically the hindwing patterns could function in three ways: [1] in mate selection, [2] as a means of deflecting a predator's first strike from the vulnerable body to the less vulnerable wing and [3] as a way of intimidating a predator that has attacked a resting moth. In fact underwing moths can pair in complete darkness and there is no suggestion that hindwing coloration plays a part in finding a mate, but we are left with two rather different possibilities. One way of deciding between them is to observe how underwing moths and predatory birds interact.

Many underwing moths bear beak marks or tears on the wings which indicate that they have escaped the attacks of birds. By observing American blue jays *Cyanocitta cristata* attacking moths released into an aviary, three different types of wing damage were recognized and related to different types of attack by the jays. In the first there is a tear in one wing which results when a bird has attacked, and lost, a flying moth; the tear is nearly always in the hindwing

Interpreting damage to a moth's wings inflicted by birds. In this North American underwing moth, there are symmetrical tears on the fore- and hindwing suggesting that the moth was able to break free of the bird's grip and escape.

which shows that the bird's attention was focused on it while making the attack. This suggests that the colour pattern of the hindwing deflects a predator's attack from the vulnerable body. In the second type of wing damage there are matching tears in the forewing and hindwing: this occurs when a bird attacks a resting moth because the wings are folded together at rest. Such moths break free from the bird's grip and escape, and the colour pattern of the hindwing can neither deflect nor startle the bird at this stage. Thirdly, there are moths with a clear beak mark on the forewing and occasionally on the hindwing as well but without any tear in the wing. These moths evidently startle by exposing the hindwings so that the bird weakens its hold momentarily and the moth escapes, not with a torn wing but with a beak mark.

This information was used to interpret the types of wing damage found in wild *Catocala* moths. In a sample of 73 bird-damaged underwings collected in a light trap in Massachusetts, 18 had a beak mark but no tear in the wing, and were considered to have escaped death by startling the predator. This is a high enough proportion for us to assume that predators can be intimidated by brightly-coloured hindwings. Observations on moths offered to blue jays show that birds were startled, but that they can learn not to be startled if they repeatedly encounter individuals with the same hindwing coloration. They learn what to expect, but as soon as they come across a moth with a different hindwing coloration they are again startled and may release the moth.

We can now return to the question of why, in the samples of underwing moths collected at four locations in North America, the frequencies of different sorts of hindwing coloration remain the same at different places. It seems that not only do conspicuous hindwings deflect or startle predators, but that because there are various sorts of hindwing coloration the chances of a

The eyed hawk-moth in the normal resting position on a tree trunk (left).

The moth intimidates its attacker by exposing large eye-spots on the hindwings (right).

predator learning what to expect are reduced. For example, the advantage of the black, white-banded hindwings in group 1 is not so much in the coloration itself but in the fact that it is different from other groups. There is a lot more to learn about flash coloration in moths; the difficulty is in separating the advantages of deflecting a predator's attack from those of startling the predator.

Mantids are predators of other insects but they are also attacked by birds, lizards and even certain large insects like hunting wasps. When in the normal resting position the majority of species of mantids are well camouflaged against the background; many look like leaves or flowers. But when attacked they defend themselves in a variety of ways. Many species have brightly-coloured hindwings and coloured marks on the forelegs which are exposed and held in a threatening posture when the mantid is attacked. This behaviour is known to intimidate birds. In one African species, *Idolium diabolicum*, the

forelegs are brightly-coloured and swollen, and there is a report of an individual rearing up and frightening away a monkey that had approached it. Flash coloration on the legs and hindwings of mantids is undoubtedly intimidating and, interestingly enough, each species not only has a different coloration but a distinctive threat display. This reduces the chances of a predator learning by experience that the mantids are simply bluffing and are not really dangerous. Most mantids are tropical; there are many species, few are especially common, and so a specialist mantid-eater would have to become familiar with a variety of colour patterns and threat displays in order to overcome intimidation.

Intimidating eye-spots

When at rest on a tree trunk or fence the European eyed hawk-moth *Smerinthus ocellata* looks like a dead and withered leaf. If a resting moth is gently prodded it opens the forewings and exposes conspicuous eye-spots on the hindwings, and at the same time rocks forward and backward in what appears to be a threatening display. The sudden appearance of two 'eyes' which move menacingly towards the observer is intimidating to a human and presumably has the same effect on predatory birds. Unlike underwing moths, the eyed hawk makes no attempt to fly away and relies entirely on intimidation as a means of protection. Indeed eyed hawks, like many large-bodied moths, are unable to take flight until they have shivered for several minutes and warmed up their muscles to a temperature that makes flight physiologically possible. Large eye-spots occur in many other moths and butterflies. Most species possessing eye-spots are palatable and rely on camouflage for primary defence. The eye-spots are normally hidden from view while the insect is at rest and are exposed only when there is danger.

The Eurasian peacock butterfly *Inachis io* has a conspicuous eye-spot on the upperside of each of the four wings. The spots are clearly visible when the butterfly perches on and feeds from a flower or when it is sunning itself on a patch of ground, but when inactive the wings are held together over the back and the upperside is completely hidden by the black underside. If a resting and inactive peacock is disturbed it performs a protective display which makes full use of the eye-spots. The wings are opened out, exposing the forewing spots, and the forewings moved forward so that the hindwing spots are fully visible. The movement forward of the forewings is accompanied by a slight hissing sound produced by rubbing together certain special veins on the forewing and hindwing. The wings are then closed together again and the whole sequence repeated several times. The butterfly orients itself in the direction of the disturbance. If a peacock is put into an aviary containing a bird it settles and performs its rhythmic display, constantly re-orienting itself in the direc-

tion of the bird. The display and the appearing and disappearing eye-spots are undoubtedly intimidating.

In an experiment to test the effectiveness of the peacock's anti-predator display, butterflies with their eye-spots removed (the scales on the wings which produce a butterfly's colourful pattern can easily be rubbed off) and normal butterflies were offered to captive yellow hammers *Emberiza citrinella*. The normal butterflies were four times more successful in intimidating the birds than those butterflies without eye-spots, but the birds soon learnt to overcome their fear of the display and after a little while attacked and ate both sorts of butterfly. The experiment shows that a bird finding a single peacock might leave it alone but would soon become habituated to the display and the eye-spots if there were repeated encounters over a short period of time. This, however, is unlikely to happen in the wild.

In the peacock, and in many other species, the protective display is performed only when the insect is disturbed from a camouflaged resting position and when escape by taking flight is impossible. In sunny weather when it is active, the eye-spots are clearly visible but no display is given; if attacked in these circumstances it escapes by flying away. Indeed it is likely that the eye-spots also serve to deflect the predator's attention from the vulnerable body.

Similar experiments involving moths of the family Saturniidae show that exposure of eye-spots accompanied by a protective display can initially intimidate birds but after repeated encounters the birds learn that there is no danger and eat the moths without hesitation. (This is the opposite to encounters with warningly coloured, unpalatable insects, which will be described in the next chapter. With these, repeated experience reinforces earlier warnings and eventually such insects are left alone.) There has been a good deal of speculation about why eye-spots feature so much in protective displays. One idea is

The extraordinary eye-spot display of the South American frog *Physalaemus nattereri*. Its backside is raised and the attacker is frightened away. Should the attacker proceed, it discovers that the frog produces an unpleasant secretion from glands located in the region of the eye-spot.

that they resemble the eyes of vertebrate animals. Small birds have an innate fear of large eyes, associating them, it is believed, with their own predators, such as owls, snakes and carnivorous mammals.

Intimidating eye-spots are not only found in adult moths and butterflies. The caterpillars of several species have them, and in caterpillars there are nearly always two, positioned on the top of the head end, so that the caterpillar seems to resemble a tiny snake. In some species the 'eyes' are large and rather ludicrous, so that the resemblance appears to be to a cartoon-type snake rather than a real one. But in others the resemblance is far more convincing, with the 'head' having a bony, angular shape, and the caterpillar's skin marked with what look

like reptilian scales. When disturbed, such caterpillars engage in a display which looks very like the behaviour of a snake. The most famous of the mock-snake displays is by the sphinx caterpillar *Leucorampha* of Central America. When at rest it looks like a twig. But when it is disturbed it transforms itself: the front end drops off the branch and inflates into a triangular snake head complete with realistic eyes. Also, the caterpillar will strike accurately at any object that touches it, which enhances the display. They certainly have a snake-like appearance to a human observer, but are birds likely to see them in the same way, or, even if they do, would they be fooled when the caterpillars are so small? It has been suggested that birds cannot discriminate size very well,

Primary defence. The South American bush-cricket, *Tanusia brullaei,* looks exactly like a leaf.

and that, to them, a snake-in-miniature is as intimidating as a real one, but whether this is true remains to be proved.

Eye-spots are also found in other insects, and even in some vertebrates. A species of tropical bush-cricket, *Tanusia brullaei*, normally camouflaged as a dead leaf, will, if disturbed, open the leaf-like forewings, and erect over its body the hindwings which carry a pair of enormous blackish eye-spots with an uncanny three-dimensional effect to them. A South American species of frog, *Physalaemus nattereri*, has a large and dramatic pair of eye-spots on its rump. If provoked it turns to face away from the attacker, and lifts its rump into the air to display the eye-spots. It has not been shown experimentally that, in these instances, the eye-spots are effective in intimidating predators, but, extrapolating from the results with eye-spots in butterflies, it is reasonable to assume that they could function in this way.

Aggressive displays

The protective display of the peacock butterfly and other Lepidoptera with large eye-spots could be described as aggressive, but it is only bluff and a persistent and experienced predator is unlikely to be fooled. Countless other animals, with or without special intimidating markings, attempt to escape predators by using aggressive displays which are nothing more than bluff.

The black, spiny caterpillars of the peacock butterfly feed in great clusters on stinging nettles. If disturbed they repeatedly jerk their bodies from side to side in unison. An entire family performs together, and the effect is dramatic and off-putting. The selective advantage of this behaviour for the individual is that if the predator is intimidated they are all left alone; the disadvantage is that if the predator is not fooled it is an easy matter to kill the entire family. It must be assumed that predators are usually intimidated and only occasionally overcome their fear and attack the

Secondary defence. The wings are opened to display the eye-spots.

Overleaf By clustering together, a family of caterpillars of the European lackey moth *Malacosoma neustria* presents an intimidating spectacle for any predator.

caterpillars, and that collective behaviour is more effective for the survival of each individual than a response given by only one caterpillar.

Collective aggressive displays are common in animals that live in groups. Gulls prey upon the eggs and nestlings of terns. If a large gull enters the airspace of a colony of nesting terns, the terns rise up into the air and mob the intruder until it leaves. Why do they do this? Would it not be better to sit tight? No, because by doing nothing one or more tern nests would be destroyed by the gull, which can drive the much smaller tern away. Collective defence often results in a failed mission for the predator. Intimidation of this sort is remarkably effective: gulls cannot harm flying terns; the terns are too agile for that, and the confusion resulting from a collective chase means that more often than not the predator is foiled.

Hole-nesting birds, including the European blue tit, hiss like a snake if disturbed while incubating eggs and small young; they may also jerk their wings in a threatening manner and puff out their plumage which adds to their bluff and intimidates predators. Young, but well-grown, long-eared owls *Asio otus* fluff out their feathers and face the intruder in a menacing way: they look enormous and, if the predator persists with an attack, the owl retaliates with its talons. When attacked puffer fish, of which there are a number of species, inflate the stomach with air or water. Most puffer fish have spines on the skin, and, when fully inflated, the spines become erect and the fish has the appearance of a prickly football. If the stomach is inflated with air the fish floats upside down and only deflates itself and swims away when the danger has passed. These fish are found in all tropical seas, and in some rivers. The side-striped chameleon *Chameleo bitaeniatus*, which is normally a camouflaged pale green or brown, suddenly darkens when faced with danger; presumably

The female of the Australian mountain grasshopper, *Acripeza reticulata*, has no hindwings and cannot fly. Until provoked, she conceals herself by holding her forewings over her body. If this should fail to protect her, she flicks her wings forward and displays a brilliantly coloured abdomen.

A young long-eared owl bluffing its way out of a difficult situation.

this puts the predator off what had been a harmless looking green or brown animal, but chameleons also darken when they come face to face with each other, which suggests that colour change also plays a part in territorial defence.

Many animals exhibit aggressive bluff when under threat, and it must be assumed that such bluff is often effective in putting off the predator. A cornered animal does not give up readily and, even if it cannot really fight for its life, it makes a pretence that this is what it is going to do.

Making a stink

Chemical warfare, especially as a means of defence rather than offence, is common in the animal kingdom. Most of the species that produce unpleasant smelling or tasting secretions are warningly coloured to advertise their distastefulness, but some are camouflaged and use repellent secretions only as a secondary line of defence. Many species synthesize

their own chemical compounds, others obtain them from plants, or even from animals they have eaten.

In the last twenty years chemists and biologists have successfully collaborated in research projects that have provided a wealth of fascinating information about chemical defence in animals. Making a stink is perhaps one of the most effective ways of secondary defence for a cornered animal. A spectacular example occurs in the aptly named bombardier beetles *Brachinus* which are found in several parts of the world. They have dark blue-black wing-cases, and an orange head, thorax and legs, and are probably warningly coloured. A bombardier beetle has an internal chamber which acts as a reservoir for chemicals which it produces: hydroquinones and hydrogen peroxide. When attacked by a predator the liquid is forced into what is called an explosion chamber containing special enzymes. The enzymes convert the hydroquinones into unpleasant-smelling quinones and the hydrogen

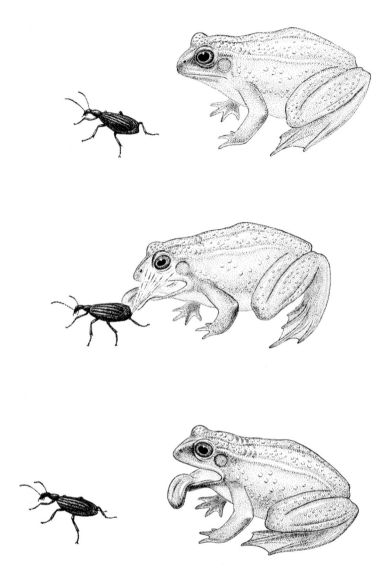

defeats the bombardier beetle. If a beetle is dropped into such a spider's web, the spider gently wraps it up in silk without stimulating the release of spray. The spider then bites the beetle which immediately fires off its spray, but with little effect, for the silk wrapping prevents the liquid from travelling.

Larvae of the sawfly, *Neodiprion sertifer*, feed on the leaves of pine trees. When startled, a larva will emit a drop of fluid from its mouth and daub it over its attacker. The fluid smells like pine resin and chemical analysis has shown that it is indeed pine resin. Resin is produced by pine trees for the very purpose of deterring insects from feeding on them. Most insects cannot cope with pine resin, but a few have become adapted to do so, this sawfly larva being one of them. The resin is taken up by the larva as it feeds on leaves, and the fluid stored in a pair of pouches in the gut. When daubed onto an insect attacker the resin acts as a powerful deterrent. Hence, in this species, a compound produced by a plant to ward off insects is taken up and used as part of one insect's own defences.

Even more bizarre is the case of the flightless grasshopper, *Romalea microptera*. This species produces a frothy fluid from the front pair of thoracic respiratory openings (spiracles), the froth itself being generated by mixing respiratory air with the liquid. The secretions contain several different chemical compounds including 2,5-dichlorophenol. This chemical is a constituent of a herbicide commonly used in the area of Florida where the grasshoppers occur. It seems, then, that the grasshopper has adopted a man-made herbicide and added it to its armoury of chemical defence – an ingenious piece of opportunism.

A frog approaches a bombardier beetle.
The beetle squirts a hot spray of unpleasant liquid into the frog's face.
The gasping frog sits back and the beetle escapes.

peroxide to water and oxygen. The pressure of oxygen gas causes the beetle to eject an explosive spray from its anus which can be pointed in the direction of the predator. The spray is not only nasty-smelling but is also hot because of the chemical reactions which have just taken place. This should be enough to intimidate even the most determined of predators, but orb-weaving spiders *Argiope* have developed a strategy that

All sorts of animals are now known to produce chemical secretions which make a nasty stink and put off an attacking predator. One of the most widespread compounds is hydrogen cyanide, produced in animals as different as millipedes and butterflies, and well known as an effective poison. Yet predators overcome the effects of unpleasant stinks and foul tastes and successfully tackle prey. It is a question of ploy and counter-ploy, in which each 'move' is part of the evolutionary battle between eater and eaten.

Making a noise and sounding a warning

One of the most effective ways of intimidating a person is to shout; this results often, but not always, in a retreat from the position adopted or at least a backing down from an incipient confrontation. If a group of monkeys is disturbed they howl, yelp or bark as they retreat or stand their ground. Even to the human observer this is intimidating and can be mildly frightening, and it certainly reminds us of our own behaviour when we are taken unawares or startled.

Terns mobbing a marauding gull make a lot of noise which adds to the effectiveness of their behaviour in driving off the intruder. Many species produce a sound at the same time as they use a protective display or make a stink: examples are the protective display of the peacock butterfly described earlier in this chapter, and even the flightless grasshopper, *Romalea microptera*, makes a hissing noise as it produces its unpleasant froth.

Visitors to the tropics often express concern at the prospect of encountering dangerous snakes, yet it is possible to spend a long time in a place where snakes are common and not catch sight of one. One reason for this is that most snakes are well camouflaged; another is that they are apt to move quietly away from a source of disturbance. In the tropics the best way to find a snake is to listen for a flock of birds chinking, chattering and chucking in a bush or tree, and then look carefully from a distance: almost certainly a snake will be revealed. Birds mob snakes, but whether they fear them we do not know. It must be assumed that

the snake is first found by one bird which immediately makes a noise and attracts other birds, of a variety of species, to the vicinity. Snakes rely on stealth and concealment when attempting to stalk a bird; a snake that has been discovered stands no chance of a successful kill.

But would it not be better for the individual bird that has found a snake to keep quiet and simply take evasive action without alerting others of the danger? In other words should not an individual be selfish and simply look after itself? Apparently not, because an isolated bird is in greater danger than a group; by warning others the chances are that the snake will be thwarted and will move away from the area. After all the isolated bird that has seen the snake may have been the potential victim but by sounding the alarm others are warned and proceed to mob the snake until it goes away or until the birds are satisfied that there is no further danger. Mobbing therefore intimidates the snake; it is exposed for what it is and has no chance of being stealthy. In the end it gives up and goes away.

Birds also mob owls, but they do this in daylight when the owl is unlikely to be looking for a victim. If you watch a group of small birds surrounding and mobbing a tawny owl sitting quietly in a tree you can see what appears to be blatant intimidation. The owl seems uneasy and uncomfortable and eventually moves away, followed for a short distance by the mobbing birds. The likely explanation for this is that tawny owls hunt at dusk for roosting birds, and these birds often use the same roosting places night after night. It is therefore in the interests of every individual bird to clear the area of a dangerous owl before darkness falls.

Some species appear to use noise in a deceptive way. As already mentioned, the blue tit, when nesting, will make a hissing noise as part of its aggressive display. To a human ear this sounds like the hiss of a snake, but we should be cautious in assuming that it has any such effect on the bird's predators. Groups of termites, when approached, rattle dead leaves with their bodies, making a sound which suggests that a large animal is hidden in the vegetation. Certain night-flying moths, when pursued by a bat, are able to emit high-pitched noises of the same kind that the bat uses in locating its prey. These noises which the bats produce act as a sort of 'radar': the bat picks up the echoes with highly sensitive ears, and uses the echo pattern to follow the prey's movements. But when faced with a moth sending out the same sort of noises, the bat's 'radar' is disrupted, and it has difficulty in catching the moth.

The anti-predator strategies of aphids

Aphids are abundant insects and there are many species: there may be up to 5000 million of them feeding from vegetation in just one hectare of countryside. They extract sugar and nitrogenous compounds in solution from the young stems and leaves with their specially modified mouthparts known as stylets. Some species, like the black bean aphid *Aphis fabae*, which feed on beans, sugar beet and other crops, occur in dense aggregations and are conspicuous; others, like the lime aphid *Eucallipterus tiliae*, which feed on the leaves of lime trees, are much more dispersed but still abundant. Aphids are preyed upon by a wide variety of animals, particularly the adults and larvae of ladybirds, and the larvae of lacewings and many hoverflies, as well as small birds like the blue tit. At first sight it seems as if aphids, unlike most animals, make little attempt to evade predators: they sit still and apparently allow themselves to be eaten.

Many species of aphids are green and are presumed to be camouflaged against a leafy background; others are black and easy to see, or whitish, and do not appear camouflaged. A few are warningly coloured: *Aphis nerii* is bright yellow-orange; it feeds on oleander *Nerium oleander*, and other related plants rich in poisonous compounds known as cardenolides which the aphids store thereby making themselves unpalatable to predators. Such secondary defences as aphids do have involve both escape and intimidation. Most species face downwards and since insect predators tend to walk up a stem or leaf they meet the aphids head-on which gives the aphids a better chance of taking evasive action. When attacked an aphid kicks itself free, or at least tries to do so, or walks away, having first withdrawn its stylet from the plant. Some aphids drop to the ground which is hazardous as death from starvation is almost inevitable. Towards the rear of an aphid there is a pair of tube-like structures called siphunculi which produce an oily, repellent secretion. The release of the secretion by one aphid can elicit escape behaviour in those nearby. Many aphids are tended by ants which feed on the excreted honeydew and in return deter or remove predatory insects. But despite these anti-predator strategies aphids seem to do little to protect themselves and are largely at the mercy of predators that move freely among them. If you watch ladybird larvae feeding among dense clusters of aphids it is easy to form the impression that they meet little resistance.

In most aphids, sexually reproducing males and females occur chiefly in autumn; these produce the overwintering eggs. In spring and summer reproduction is by parthenogenesis: there is no mating, females producing daughters genetically identical to themselves. The rate of reproduction is so rapid that a single female could theoretically give rise to six hundred thousand million offspring in a season. An aphid weighs about 1 milligram and a spring female therefore has the potential of producing six hundred thousand kilograms of offspring by the end of the year, all of them identical to herself. From the evolutionary point of view aphids produced in this way are all the same individual. Hence when a predator attacks and eats

When a puffer fish is attacked, it inflates itself and becomes a prickly football. It remains like this until the danger has passed, and then deflates itself and swims away.

an aphid it is merely removing a tiny piece from one gigantic individual; it does about as much damage as a caterpillar does when it chews the leaves of a large tree. In the evolutionary sense, and this is what matters when we are speaking about the adaptive significance of anti-predator strategies, individual aphids are rarely if ever killed by predators: parts of the individual always survive and reproduce and the effect of predators on the total genetic unit is minimal. Aphids are therefore a special case. It now becomes clear why their more conventional anti-predator strategies are so poorly developed: they can afford to be eaten.

Intimidation under water

Much of this chapter is about animals that live on land. This is because aquatic species have been studied in less detail, and because land animals, especially insects and birds, are more satisfactory to use in experiments. There are many uncertainties about the meanings of what might be intimidating behaviour in aquatic animals.

For example, four different families of bony fish, as well as certain rays and torpedoes (cartilaginous fish), have, independently of each other, evolved the ability to produce electric discharges. The freshwater Mormyridae of African rivers and lakes produce electric discharges which are apparently used in signalling to each other. The Nile perch *Gymnarchus niloticus* puts out discharges as a way of detecting food. There are several species of electric rays, belonging to the family Torpedinidae, which are found in all tropical and subtropical seas. The Mediterranean and east Atlantic species, *Torpedo narke*, stuns its prey with discharges, but can also stun a potential predator, a fine example of intimidation. When at rest on the sea bed this electric ray blends well with its surroundings except for five conspicuous, black-ringed blue spots, which resemble eyes and which perhaps are a warning to predators of its ability to produce an electric discharge. Electric discharges seem to have several functions and a great deal more research is needed before we know all the answers.

Much the same is true of luminous organs. Most fish, crustaceans and cephalopods living below a hundred fathoms are luminescent and some have lights that can be flashed on and off. Many species generate their own light but in some these lights are produced by luminous bacteria contained in special structures in the skin. Various explanations have been put forward for the function of luminous organs in deep-sea animals. In groups as different as shrimps, cuttlefish and hatchet-fish, the luminescence is on the underside and aids in breaking up the shadowy silhouette of the animals when viewed from below: it is thus a form of disruptive coloration. Luminous organs may also help in locating mates or in finding prey. Is it possible that, at least in some species, they are also used to intimidate predators in much the same way as the eyespots on the wings of butterflies and moths? This is certainly a possibility. One parallel exists in a species of deep-sea octopus which, when disturbed, emits a cloud of luminous fluid in place of the cloud of 'ink' produced by other octopuses.

8   Warning coloration

If an animal is red, orange, yellow or white with bold black stripes, bands or spots the chances are that it is dangerous or unpalatable to predators. The protective strategy of warningly coloured animals is the opposite to that of camouflaged animals. These are hard to see and tend to keep still. Warningly coloured animals make little attempt to conceal themselves and, with a few exceptions, are active and conspicuous by day. Predators learn to avoid them just as they learn to find camouflaged ones.

Bold yet simple warning coloration advertises one of the following: the animal can sting, irritate, inflict a poisonous or painful bite, produce a nasty smell, taste foul if eaten, or is poisonous. But it must not be assumed that bright colours and bold patterns always signify danger and unpalatability. Many birds and butterflies are brilliantly coloured but the coloration functions in courtship and sexual display. This is usually the case when the bright coloration is found in one sex only, in most species in the male. Many fish of shallow tropical seas, particularly those that inhabit coral reefs, are likewise brilliantly coloured but so are their surroundings and it is possible that in reality they are well camouflaged. Remember, too, that a fish may look brilliantly coloured to us but for a predatory fish with a different sort of vision it may be difficult to see. There are also many harmless and palatable animals that mimic the colours and patterns of noxious species and so gain protection from predators by deception. These will be described in the next two chapters.

Warning coloration is an animal's primary defence against predators; once attacked its secondary defence is to sting, irritate, bite, smell nasty, taste foul or to be poisonous, although these last two tactics look as if they have disadvantages for the predator must attempt to eat the animal to discover its defence. Dangerous and unpalatable animals are by no means fully protected. As always there are predators able to overcome protective armoury and, as with camouflage, no animal is entirely safe. Indeed the selective advantage of warning coloration may be quite small but enough to account for its evolution in a wide array of species.

Stings and irritations

Social wasps *Vespula*, *Polistes* and related genera, bumblebees *Bombus* and the honeybee *Apis mellifera* are the most familiar insects of temperate regions which we immediately recognize as being able to inflict a painful sting. *Vespula wasps* are yellow and black, a common combination in warningly coloured insects, but *Polistes*, honeybees, and bumblebees are simply wasp- or bee-like in appearance and are recognized as such by predators. The sting is a modified ovipositor (egg-laying structure) and so only females are able to cause pain. When a bee stings, the sting with venom sacs attached remains in the wound and the bee dies. It is the non-reproducing worker honeybee that stings and hence a death makes evolutionary sense: because it cannot produce offspring the worker is really part of the reproducing queen bee and forfeits its life in defence of the colony.

A few predators are able to cope with bees and wasps. Birds like the bee-eaters of the family Meropidae catch bees, take them to a perch, and by carefully hitting them against the perch squeeze out the venom and render the bees harmless before eating them.

Many fish have protective spines, particularly on the gill covers and dorsal fins; in some species there are skin glands containing poison which is injected into the wound inflicted by the spines. Some but not all fish capable of injecting poison are brightly coloured and the coloration may serve as a warning to predators.

Among amphibians there are several species which produce a distasteful or highly poisonous secretion from their moist skins. In some species the secretion is produced when the animal is disturbed and is evident as a frothy liquid, while in others the skin appears to carry the secretion at all times. Whereas amphibians are generally coloured green or brown, and appear to be camouflaged, a number of those which are distasteful or poisonous are also brightly coloured. Their skins may be brilliant red, blue or yellow, or some combination of these colours, with black often featuring in the pattern. An excellent example is the South American tree-frog, *Dendrobates histrionicus*, one species whose skin was reported to be the source of the poison used on the arrow-tips of Amazonian Indians. Another instance of this is seen in the European fire salamander *Salamandra salamandra* which is patterned with

Dendrobates histrionicus, a South American arrow-poison frog, one of a number of species whose skins were reported to be the source of the poison used on the arrow-tips of Amazonian Indians.

irregular patches of black and vivid yellow. A milky discharge which is highly poisonous is produced by the skin of this salamander.

Hairy caterpillars are often brightly coloured and, even if they are not, they are easy to see as they do not attempt to conceal themselves. When handled the hairs are dislodged and can produce a rash on the human skin. Caterpillars of the brown-tail moth *Euproctis chrysorrhoea* and the yellow-tail moth *Euproctis similis,* both locally abundant in Europe, are gaudily coloured and easy to see. Children pick them up and sometimes get a rash, but birds leave them alone, except cuckoos which seem to be able to eat them with no ill-effects. Adult brown- and yellow-tail moths are shiny white with a tuft of brown or yellow hairs at the tip of the abdomen. The hairs are dislodged if the moths are handled and can cause a nasty rash. Both species lay clusters of eggs on twigs and cover them with protective hairs from the abdomen. Although not brightly-coloured the hairy brown caterpillars of the garden tiger moth *Arctia caja,* the buff ermine *Spilosoma luteum,* and the white ermine *Spilosoma lubricipeda,* all common European species, are usually ignored by birds. In the tropics the caterpillars of several groups of moths, especially the Limacodidae, produce a stinging sensation when hand-

A change of tactics. The young caterpillar of the alder moth *Apatele alni* resembles a bird dropping. When fully grown, it is yellow and black and, possibly, warningly coloured.

led; some species are brightly coloured but others are believed to be camouflaged; all have spines and hairs that cause the irritation.

The two-spot ladybird *Adalia bipunctata* and the seven-spot ladybird *Coccinella septempunctata* are both red with black spots. If offered to captive birds they are ignored or rejected, but swifts sometimes catch and eat them in the air where, it seems, warning coloration in small beetles cannot be seen. Ladybirds are sometimes reported as 'biting' people and causing a skin rash. When prodded a ladybird produces a yellow secretion from joints in its legs. The secretion, an alkaloid called coccinellin, has a nasty smell and can cause a rash which is presumably the origin of reports of ladybirds 'biting'.

Irritations can also result from squashing certain insects on the skin. The body fluids of some rove beetles of the family Staphylinidae can produce a most unpleasant rash. One African species, a member of the genus *Paederus*, is nicknamed 'Nairobi eye' because it causes a nasty swelling on the delicate skin surrounding the eye if its body fluids are accidentally rubbed on the face. *Paederus* is orange and black and if one should happen to alight on the skin or clothing it should be flicked away, not squashed. This a wise policy for dealing with any brightly coloured insect.

Poisonous bites

Bugs of the family Reduviidae have piercing and sucking mouthparts which they use to attack their prey. Some species are warningly coloured, usually orange and black or red and black. If handled they can pierce the skin and cause an unpleasant stinging sensation. In these bugs the same tactic is used in both offence and defence.

The same is true of poisonous snakes, most species of which, however, are camouflaged, not warningly coloured. But the South American coral snakes, so named because of their gaudy coloration, are banded with red, black and yellow. They are active during the day and, unlike most snakes, do not conceal themselves. The coral snakes comprise about 50 species, all of which look remarkably similar. Two families are represented, the Elaphidae which are deadly poisonous, and the Colubridae which contains both slightly poisonous and harmless species. The deadly poisonous elaphids rarely bite in defence but do so to kill prey. On the other hand the slightly poisonous colubrids defend themselves from predators by biting. Because of the similar coloration predators have difficulty in distinguishing the different types of snake and it is believed that both the deadly poisonous elaphids and the harmless colubrids are mimics of the slightly poisonous and aggressive colubrids.

Nasty smells and foul tastes

When theatened, the black and white North American skunk *Spilogale putorius* stands on its forelegs and advances towards the attacker. If the attacker does not retreat the skunk squirts an unpleasant-smelling fluid from its anus. The striking black and white colours of the skunk are in contrast to the dull browns and counter-shading of most mammals, and probably act as warning coloration. If a butterfly of the predominantly African family Acraeidae is handled it produces copious quantities of yellowish foam from thoracic glands. The foam smells like hydrogen cyanide, and in one species, *Acraea encedon*, it has been confirmed that hydrogen cyanide is present in the foam, although there are other chemical compounds as well. There are many species of Acraeidae; most are orange and black or black and white, or some similar combination of bold colours. The forest species in particular fly slowly and deliberately and rarely take evasive action when threatened.

Warning coloration is common in tropical butterflies but rather unusual in temperate species. There are, however, several warningly coloured European moths. The day-flying, scarlet and black cinnabar moth *Tyria jacobaeae* and the boldly marked red, white, black and brown garden tiger moth *Arctia caja*, a night-flyer, contain in their tissues chemical compounds derived from the plants on which their caterpillars feed; they also synthesize their own poisonous compounds. Both species are highly unpalatable and are rejected by birds on sight. As already mentioned, the caterpillars of the garden tiger moth have irritant hairs; those of the cinnabar are yellow, banded with black; they feed unconcealed in big family clusters, and are ignored by most predators. The fully-grown caterpillar of the European alder moth *Apatele alni* is similarly banded with black and yellow, but the young caterpillars look like fresh bird droppings. In this species the young caterpillar's primary defence is looking like something inedible, while the fully-grown caterpillar is apparently warningly coloured. It would be interesting to know if the caterpillars also change from being palatable to being unpalatable as they grow larger. The brilliant scarlet and black burnet moths *Zygaena* synthesize hydrogen cyanide; their tissues also contain histamines. Burnets themselves are remarkably resistant to cyanide poisoning, and the caterpillars are able to feed on plants containing the compound in their leaves.

Monarch butterflies of the family Danaidae are found in North and South America, in Africa, Southeast Asia and Australasia. Many species are orange and black or black and white. The caterpillars feed on milkweeds of the family Asclepiadaceae from which they absorb and store cardenolides. These poisons are stored through the pupa stage and are present in the tissues of the adult butterflies. Captive blue jays vomit after being fed on the North American monarch *Danaus plexippus*.

Chemical analysis of the African monarchs (sometimes called the African queen) *Danaus chrysippus* has shown that there is variation in the amounts of

The yellow and black fire salamander *Salamandra salamandra* is warningly coloured and unpalatable to most predators.

Overleaf This Arizona coral snake *Micruroides euryxanthus* is one of about 50 species of brightly coloured coral snakes. Its vivid markings warn that it is highly poisonous.

cardenolides stored by the caterpillars. Some species of milkweeds which the African monarch feeds on lack cardenolides, some have low concentrations, while others are rich in the compounds. The concentration of cardenolides in the food-plant partly determines the palatability of the butterflies. But there is more to it than this: even when reared on the same food-plants the African monarch is a rather inconsistent storer of cardenolides. This suggests that, besides variation in palatability resulting from the kind of milkweed the caterpillar has fed upon, there is also a genetic factor which makes some individuals better storers than others. Indeed wherever the African monarch is found it seems that there are both palatable individuals and a whole range of unpalatable individuals, from the mildly distasteful to the extremely distasteful. The butterflies look identical to each other, so a predator could not be expected to tell them apart except by taste. This becomes important when we discuss the mimics of the African monarch in Chapter 10.

Most swallowtail butterflies are palatable but in various parts of the world there are species whose caterpillars feed on the leaves of pipe-vines of the family Aristolochiaceae. Some, but not all, of these swallowtails contain poisonous aristolochic acids in their tissues which are absorbed by the caterpillars from the food-plant. The common pipe-vine swallowtail *Battus philenor* of eastern North America contains aristolochic acids, and experiments show that birds quickly learn to avoid eating them. The largest butterfly in Africa, *Papilio antimachus*, also a swallowtail although it does not look like one, is rich in cardenolides, and must be a very poisonous butterfly. It is relatively rare and its caterpillars and food-plant have never been found, but the presence of cardenolides in the adult butterfly suggests that the search should concentrate on those plants known to contain these compounds.

Are some plants warningly coloured?

The cardenolides present in milkweeds are part of the defensive strategy evolved by the plants to inhibit and reduce consumption of their leaves, stems and flowers. A few plant-feeders like monarchs not only overcome this line of defence but also make use of the poisons for their own defence. Most plants defend themselves by either chemical or physical means: warning smells and tastes, and tough, prickly or hairy textures to leaves and stems. A few species seem to have evolved warning coloration, or what looks like warning coloration, but it must be admitted that experimental proof is lacking and the evidence is circumstantial.

Fly agaric *Amanita muscaria* is a poisonous toadstool found in woods in northern Europe. The brilliant scarlet cap and the white gills and stalks are at first enclosed in a white veil, but this soon falls apart leaving patches sticking to the top of the cap. This gives a striking speckled coloration making the toadstool 'look poisonous' to an extent that few people would try eating it. On the other hand the related deadly poisonous death cap *Amanita phalloides* is greenish- or yellowish-white and not obviously warningly coloured, nor is the equally poisonous destroying angel *Amanita virosa*.

The toadstool is the reproductive part of a fungus, and the tiny spores it produces are dispersed by wind. If a toadstool is eaten its delicate spores are very likely to be destroyed, so that anything which reduces the chances of the toadstool being eaten would be advantageous to the fungus. Fruits, on the other hand, are especially designed to be eaten. The succulent outer flesh encloses hard seeds, which resist the digestive processes and, after passing through an animal's intestine, can germinate. Being eaten is advantageous because in the process the seeds are dispersed and the plant may then become established in a new place, at some distance from the parent plant.

Why then are some fruits poisonous, at least to man? And are the colours of these fruits — the red of woody nightshade *Solanum dulcamara* and the black of deadly nightshade *Atropa bella-donna* — warning colours, as red and black so often are in the animal kingdom? Fruits which are edible to man, such as blackberries and raspberries, also have these colours, and it seems that in general the colours serve the purpose of attracting the attention of animals which, by eating them, perform the service of distributing their seeds. The important question to be answered is whether these fruits could benefit from being selectively poisonous, so that groups of animals which are less efficient at distributing the seeds are deterred from eating the fruits. If so then the colours may serve a dual purpose — to attract some animals and to warn off others.

A few plants have brightly coloured and boldly marked leaves which suggest warning coloration. The leaves of tropical species of *Caladium* (members of the family Araceae) are variously marked with red or white. *Caladium* is poisonous, and so is the dumb cane *Dieffenbachia*, another member of the Araceae whose leaves are either yellow or green speckled with white. The difficulty is that these plants have been extensively bred for ornamental purposes and although brightly-coloured varieties occur in the wild they may be escapes from cultivation.

How predators respond to warningly coloured prey

The presence of beak marks on the wings of warningly coloured butterflies and moths shows that they are often attacked but then rejected. Observations indicate that birds quickly learn from experience to reject and ignore warningly coloured insects, and others, such as honeybees which simply look 'bee-like'. Moreover a predator that has learnt to avoid, let us say, a yellow and black cinnabar moth caterpillar also avoids yellow and black wasps. The learning process is based primarily on the experience of the individual, but there is evidence that birds can learn to a limited extent from each other. It seems also that there is an innate (ie genetic) tendency for predators to avoid certain food items which, however, is usually reinforced by trial-and-error learning. The efficacy of warning coloration, and indeed of all protective devices, depends on circumstances: a hungry predator will attack and eat a noxious animal which it would have ignored if it had not been hungry. Warning coloration signifying danger or unpalatability is certainly an effective means of defence and, as we shall see in the next two chapters, predators are further confused by the wide variety of palatable species that mimic the coloration and often the behaviour of dangerous or noxious species.

9 Mimicry

In 1862 the English naturalist Henry Walter Bates published the results of his extensive observations on the butterflies of the Amazon forests. His most important conclusion was that in these forests there are brightly coloured, apparently unpalatable butterflies flying together with strikingly similar yet unrelated species of palatable butterflies. Bates suggested that palatable species, the mimics, gained protection from predators by resemblance to unpalatable species, the models, and that the similarity between the two groups was the result of evolution by natural selection. It will be recalled that in 1859 Charles Darwin first published his book on evolution by natural selection *The Origin of Species* and that throughout the second half of the nineteenth century controversy raged about the validity of the theory, not only for plants and animals but for man himself: many people, including some eminent scientists, could not accept that the wonderful diversity of living organisms, culminating in man, had been produced by so brutal a force as natural selection. Bates' observations on mimicry provided crucial evidence in support of Darwin's theory.

Mimicry is confusing; it involves deception on a grand scale; coloured illustrations of models and mimics do not really do justice to the remarkable resemblance between them: it is necessary to see living animals in their natural surroundings to appreciate fully the extent of the deception. Most of the best examples of mimicry are found in the tropics but even in an English garden there is a confusing array of models and mimics if you know what to look for.

Mimicry is now known to occur in many groups of animals including birds, fish, snakes, amphibians, moths, beetles, bugs, flies and snails, but by far the most convincing examples are found in butterflies, especially tropical ones. A butterfly's wing is a perfect structure for an elaborate colour pattern and this is probably why by far the best examples of mimicry are found in butterflies. Since Bates' time, field studies, and more recently laboratory experiments, have unfolded how mimicry works and how it is evolved; indeed research on mimetic butterflies has probably contributed more to our understanding of Darwin's theory of evolution by natural selection than any other single line of investigation.

Kinds of mimicry

Camouflage, including the resemblance to inanimate objects like stones and bird droppings, is sometimes placed under the heading of mimicry. Indeed one very broad definition of mimicry is that it is a process in which one animal (called the operator) is unable to distinguish a second organism (the mimic) from either another organism or from part of the physical environment (the models), the consequence of which is to increase the chances of survival of the mimic. There is considerable justification for such a broad definition as this because the kinds of selection pressures involved are similar, but it is usual to speak of a resemblance between two organisms, one of which is warningly coloured and unpalatable, as mimicry. This phenomenon is now known as Batesian mimicry. This is a neat definition but as usual it is far too simple. One difficulty is deciding what is unpalatable and what is palatable. Much depends on the predator, how hungry it is, the circumstances in which it is seeking food and its tolerance to poison levels. Added to this, there may be variations in palatability between members of the same species, as already described for the African monarch butterfly. Everything suggests that there is a complete spectrum of palatability and that there are some predators that will eat anything remotely suitable provided they can find it.

Bates simply assumed that some butterflies are unpalatable. He had no information about the causes of unpalatability but in most instances his assumptions were correct. By watching the behaviour of different species he judged which were models and which were mimics, noting that the mimics were usually less abundant than the models, an important requirement because predators would otherwise soon learn that their coloration does not necessarily indicate unpalatability. Later on other butterfly watchers realized that several, often unrelated, unpalatable species may resemble one another. Thus the African monarch *Danaus chrysippus*, a cardenolide storer, and the acraeid, *Acraea encedon*, which secretes hydrogen cyanide, are similar in coloration and often occur together. Which is the model and which the mimic? The answer is that each is both model and mimic: they gain protection by looking like each other, a form of deception now

called Müllerian mimicry after the nineteenth century German zoologist Fritz Müller. It often appears to be the case that the different species in a Müllerian assemblage are not all equally unpalatable, and so there are mimics, not necessarily completely palatable, but less unpalatable than the models. Indeed there is probably no clear-cut distinction between Müllerian and Batesian mimicry: the two kinds are inextricably interwoven.

What does the predator make of this? Faced with a variety of species, some palatable, some noxious, and many in between, it can only learn from experience, and the chances are that it will err on the side of caution. Predators constantly 'test' models and mimics. If a palatable mimic is temporarily too abundant in relation to its model it is discovered and eaten. The overall outcome is a balance, maintained by selection by predators, which changes a little from time to time, but which in the long run increases the chances of survival of the mimics, whether Batesian or Müllerian.

There are also cases of the predator, or perhaps we should say the robber or parasite, assuming the role of mimic. The yellow and black death's head hawk-moth *Acherontia atropos* looks like an enormous bee or wasp. In Europe it regularly enters bee-hives and steals honey and is apparently unmolested by the bees. Species of cuckoo bees *Psithyrus* are social parasites in the nests of bumblebees *Bombus* and look like the species they utilize. Female cuckoo bees enter bumblebee nests and lay eggs; bumblebee workers feed the resulting cuckoo bee larvae, a clear case of exploitation of one species by another made possible, it seems, by mimetic resemblance.

In mimicry, similar coloration is evolved in unrelated species. The resemblances are at once striking yet superficial; they are deceptive but do not fool a skilled observer: a hoverfly that looks like a wasp still has (like all flies) only one pair of wings, not two pairs as in wasps. Mimicry is essentially deception by trickery.

Mimicry in North American butterflies

The North American monarch *Danaus plexippus* is warningly coloured and abundant, two of the requirements for a species to act as a model. It stores cardenolides absorbed by its caterpillars while feeding on milkweeds and it is recognized by birds as an inedible butterfly. The viceroy *Limenitis archippus*, which is palatable, has the same colour and markings as the monarch even though it belongs to a different family, the Nymphalidae. It occurs in the same area as the monarch but is usually much less common. Birds soon learn to reject viceroys if they have previously tried to eat monarchs. This seems a straightforward example of Batesian mimicry but as is often the case more detailed investigations have revealed complications to the story.

Samples of monarchs collected at various places in North America show that individual butterflies vary in their cardenolide content. Fifty butterflies collected in Massachusetts were offered to captive blue jays but only twelve caused the jays to vomit. This suggests that only one in three butterflies was unpalatable. The result led to the suggestion that within a species some individuals are models and some are mimics, a phenomenon now known as 'automimicry'. Under these circumstances we might expect that, at least in some places, automimics would come to outnumber models and that predators would soon learn that most monarchs are palatable. It seems, however, that if a predator receives an occasional reminder, it will avoid all butterflies of a monarch-like appearance. Some monarchs are extremely unpalatable and one, reared on the poisonous milkweed, *Asclepias humistrata*, contained such a high concentration of cardenolides that it induced vomiting in eight blue jays when shared between them. This is a very nasty butterfly and occasional encounters with such individuals are apparently sufficient for birds to avoid most monarchs and mimetic viceroys.

In Trinidad, the monarch occurs together with the related queen butterfly *Danaus gilippus* which has a similar coloration. Tests show that about 65 per cent of the monarchs but only 15 per cent of the queens are unpalatable to blue jays. This means that unpalatable monarchs and queens are Müllerian mimics of each other, that palatable queens are Batesian mimics of unpalatable monarchs, that palatable monarchs are Batesian mimics of unpalatable queens, and that in both species there is automimicry. This is the sort of situation that creates confusion not only for predators but also for anyone trying to unravel what at first sight seemed a simple mimetic association.

The pipe-vine swallowtail *Battus philenor* stores aristolochic acids absorbed by its caterpillar from the foodplant; if offered to birds the butterflies are usually rejected. The pipe-vine swallowtail is mainly black with yellow, orange, blue and white spots, chiefly on the hindwings. Its distribution in the eastern United States is from New York State to Georgia, and Florida where, however, it is relatively rare. Several species of butterflies are Batesian mimics of this common swallowtail, among them the spice-bush swallowtail *Papilio polyxenes*. In Tennessee and North Carolina the pipe-vine is about eight times as abundant as the spice-bush swallowtail and in this area the two species are very similar in coloration. But to the south in Georgia and Florida the mimic is much more common than the model; here it is more variable in coloration and a much less impressive mimic. This shows that in places where predators only occasionally encounter the model, selection on the mimic to resemble it is reduced and the mimetic resemblance is less obvious.

The red-spotted purple *Limenitis arthemis* which is closely related to the viceroy, is another mimic of the pipe-vine swallowtail. Where it occurs within the range of the pipe-vine swallowtail it is black and mimetic, but to the north of the pipe-vine swallowtail's range it is not mimetic: it has broad white bands on the upperside and underside of both pairs of wings, making it look totally different from mimetic individuals. Butterflies intermediate between the two geographical forms occur near the edge of the range of the model. Again mimetic resemblance disappears in the absence of the model. (See picture on p. 138.)

The tiger swallowtail *Papilio glaucus* is yet another mimic of the pipe-vine swallowtail. In this species the males are mainly yellow with black markings and are not mimetic. There are two forms of the female, one like the non-mimetic male, the other a black mimic of the pipe-vine swallowtail. The frequency of black females varies with the abundance of the model, reaching 93 per cent in the Great Smoky Mountains where the model is common and declining to zero in the north where the model is absent. In Florida, where the model is rare, black female mimics are rare, but in some places the correlation between the frequency of models and mimics is not so good. In Georgia, for example, 96 per cent of female tiger swallowtails are black although the pipe-vine swallowtail model is scarce. It is not known why this is so but one possibility is that, because the tiger swallowtail is migratory, the frequency of black females may change from year to year depending on the amount of migration that has taken place.

When a male butterfly mates it transfers a parcel of sperm, called a spermatophore, to the female: only one spermatophore results from each mating. Spermatophores can be dissected out of females and it has been found that, at least in some places, black female tiger swallowtails mate less often than the yellow females whose pattern is like that of the males and thus not mimetic. This suggests that males are more attracted to females of the same coloration as themselves than to black females. Thus the advantage to black females in mimicking the pipe-vine swallowtail is countered, to a small extent, by poorer mating prospects.

Swallowtail butterflies are ideal for genetic research: even the tropical species can be successfully maintained in heated greenhouses and the only problem is in getting sufficient quantities of the right food-plant for the caterpillars. They can be mated by holding each butterfly by the wings, gently pressing the tips of the abdomen together and manipulating the genitalia until the male grasps the female. This means that matings can be arranged and the genetics of polymorphic colour forms worked out.

As we shall see in the next chapter, the restriction of mimetic colour forms to one sex only, usually to the female, is quite common in butterflies. However, the mode of inheritance of the black phenotype in the tiger swallowtail is unusual, not only for butterflies, but for animals in general, for black females almost invariably give rise to black daughters and yellow females give rise to yellow daughters, no matter what male is used as a mate. How can this be explained? The most likely answer is that the gene that produces a black phenotype is on the Y chromosome which only females carry. In butterflies, as in birds, the female has the XY complement of sex-determining chromosomes and the male the XX complement. If a male tiger swallowtail mates with a black female all the female offspring will be black if the trait is transmitted by the Y chromosome: there is no way for the male's sex chromosome to shield the effects of the female's Y. It will be recalled that egg colour in cuckoos may be inherited in the same way. Another possibility is that the inheritance of the black female phenotype is by a factor in the cells transmitted from mother to daughter independently of the genes: more controlled matings of tiger swal-

A honeybee (left) feeding from a flower next to a hoverfly, *Eristalis*. The hoverfly looks and behaves like the bee and is often mistaken by predators for one.

The top row shows three species of bumblebee models. In the middle there are three colour forms of *Merodon equestris*. Along the bottom are two colour forms of *Volucella bombylans* and a specimen of *Volucella pellucens* (right).

lowtails are needed to test which of these possibilities is correct.

Bee mimics

Many of the insects in a European garden that look like bees are flies. Bees belong to the order Hymenoptera and have two pairs of wings; flies are Diptera and have only one pair. Nevertheless the resemblance between the bee models and the fly mimics is striking and deceptive.

In Europe, the hoverflies *Eristalis tenax* and *Eristalis pertinax*, and several

similar-looking species, resemble and behave like honeybees. They visit flowers for nectar and pollen and most people mistake them for bees. They are often extremely abundant, especially in late summer and autumn when *Eristalis tenax* in particular is migratory and sometimes appears in swarms. Very often the mimics far outnumber the models but presumably a sting from a bee is something a predator remembers for a long time and hence the mimicry is effective. Experiments in the United States have shown that birds and toads refuse to eat *Eristalis* if they have previously had unpleasant experiences with honeybees. The 'bees' that Samson noted as bringing forth sweetness from the carcase of a lion were presumably a species of *Eristalis*, as the larvae are scavengers and feed on decomposing organic matter. Evidently mimicry has fooled people for a long time.

Another common garden hoverfly, the large narcissus bulbfly *Merodon equestris* is an excellent mimic of bumblebees and even makes a similar buzzing sound. It is hairy and the colour and patterning of the hairs are extremely variable. The variation is inherited through a complex genetic system (which has been worked out) which produces many phenotypes. Most of these are good mimics of different species of bumblebees, each of which has a distinctive colour and pattern. The large narcissus bulbfly appears in English gardens in June and early July when worker bumblebees are most abundant. The mimics are about the same size as the worker models and considerably smaller than the large queens. The larvae of the large narcissus bulbfly feed on bulbs and the insect is sometimes regarded as a pest; it has been accidentally introduced with the bulb trade to many parts of the world, including North America.

Another hoverfly, *Volucella bombylans*, is a polymorphic mimic of at least three species of bumblebees. It is most common in woodland, but in England regularly occurs in gardens. On first consideration it appears to be a Batesian mimic, but the females lay their eggs in the nests of bumblebees where the hoverfly larvae scavenge. The resemblance to the bees might be useful to the female hoverfly in gaining entry to the nest to lay her eggs. There is no reason why mimicry should not serve both purposes.

A proliferation of wasp mimics

One way of judging if a warningly coloured insect is dangerous or unpalatable is to see how many species of insect mimic it. If there are many mimics its warning coloration is really effective as protection from predators, but if there are only a few, or none, it is probably only mildly distasteful. Using this criterion, the European and North American wasps of the genus *Vespula*, and some of their close relatives, must be among the nastiest insects in the world. These yellow and black wasps inflict a painful sting when provoked. The characteristic and easily recognized wasp coloration occurs in many harmless insects including beetles, moths, mantispids, (which belong to the order Neuroptera and should not be confused with mantids), and especially flies. We do not know if all these insects are wasp mimics but some certainly are and there is reason to suppose that a general resemblance to wasps affords protection.

In an English garden there may be twenty or more species of hoverflies which resemble wasps. All have yellowish bodies banded with black, or black bodies banded or otherwise patterned with yellow. In some years certain species are migratory and become temporarily extremely abundant giving rise to reports in the press of 'plagues of wasps'. Members of the genus *Chrysotoxum* are strikingly like *Vespula* wasps: they have the same bold coloration and are about the same size and, moreover, they sit about on flowers in a wasp-like manner.

Most of the wasp-like hoverflies are believed to be Batesian mimics, but several large species of *Volucella* lay eggs in wasps' nests. When a female enters the nest it is not molested by the wasps; eggs are laid on the papery surface of the cells containing wasp larvae and pupae. After hatching, the hoverfly larvae fall to the bottom of the nest which is where the wasps throw their dead. The larvae scavenge on the dead wasps and also enter cells containing live larvae and stimulate them to produce excrement upon which they also feed. This relationship between hoverfly and wasp is probably mutually beneficial: the hoverfly larvae obtain nourishment and the wasps are left with a cleaner nest. Mimicry in *Volucella* is thus a means of gaining access to wasps' nests. The abundant *Volucella pellucens* is black with a broad white band on the abdomen. It is not a good mimic but its larvae are repeatedly found in the nests of the common wasp *Vespula vulgaris*: possibly its bold and distinctive coloration is recognized and accepted by wasps

A *Vespula* wasp (top) can inflict a nasty sting. Many insects mimic wasps, including a hoverfly, *Chrysotoxum* (middle) and the hornet clearwing *Sesia apiformis*.

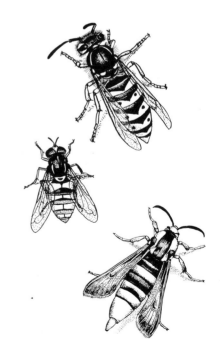

as belonging to a useful intruder; in other words instead of gaining access by deception it is recognized and allowed entry.

Several species of flies of the family Conopidae are wasp mimics, some of them more like certain solitary wasps than species of *Vespula*. Female conopids chase after and attach their eggs to the bodies of wasps and bees. The resulting larvae feed internally on the blood of the host and eventually pupate inside its body. British species of conopids have been reported as parasites of social wasps, bumblebees, honeybees and a variety of solitary wasps and bees. Unfortunately there are not sufficient records to indicate if species that resemble their hosts habitually parasitize those hosts, but if they do mimicry could be a means of approaching the host to lay eggs. Conopids, like hoverflies, feed from flowers alongside wasps and bees and may therefore also be Batesian mimics.

Moths of the family Sesiidae, often called 'clearwings', bear a resemblance to wasps of one kind or another. Unlike most other moths and butterflies, their wings have few coloured scales and are transparent. The European hornet clearwing *Sesia apiformis* is similar to a *Vespula* wasp, especially when sitting newly-emerged from the pupa on a fence or tree trunk. Superficially it is not a bit like a moth but close inspection soon reveals moth characteristics: apart from numerous details of structure it is soft and delicate, not hard and brittle like a wasp and, of course, it does not sting. Many of the smaller clearwings look like parasitic wasps of the family Ichneumonidae, most of which do not sting. Why the resemblance? It is difficult to say except that small ichneumons do not look particularly inviting as items of food and are perhaps ignored by predators.

Looking like ants

Wherever there is vegetation there are ants; in the tropics in particular there are ants everywhere and they are easily

Many spiders and small insects bear a remarkable resemblance to ants and thereby gain protection from predators or access to ant nests where they can scavenge.

the most obvious of the smaller insects. They are social, often nesting in huge colonies from which foraging parties of workers go out to seek food for the larvae. Although there is considerable variation in structure and size, all ants are distinctly ant-like, varying in colour from pale or reddish-brown to black; they are not gaudily coloured like unpalatable butterflies, moths and wasps. Some species sting or bite, others contain unpleasant compounds in their tissues. Many predators avoid ants, although there are exceptions: in the tropics, particularly, there are specialist anteating mammals like pangolins, the aardvark and the giant anteater, all of which, however, more commonly feed on termites.

Since most predators keep well clear of ants, an astonishing number of insects and spiders resemble them. The resemblance is not only in colour but in behaviour. Ants are active, fidgety, busy insects, always running around doing things and constantly making use of their antennae and legs to communicate with each other and to test possible food items. Their mimics behave in the same way.

A mimetic resemblance to ants has been independently evolved in several different families of spiders. Like all insects, ants have three pairs of legs; spiders have four pairs, but in species that mimic ants the front pair is held forward so as to look like ant antennae. These spiders also have a constricted 'waist' which enhances the resemblance. The immature stages of several species of grasshoppers, stick insects, bugs, and mantids are also ant-like.

Those insects which mimic ants only when young are presumably Batesian mimics, that is, they gain protection from predators by looking like ants. Certain of the spiders may also be Batesian mimics, but most ant mimics are directly associated with the ants themselves. They enter or live entirely in ant colonies and act as scavengers. Presumably their resemblance to ants enables them to live safely in what is potentially a hostile environment. At least eighty species of rove beetles of the family Staphylinidae live in ant nests and most are plausible ant mimics. In all there are thousands of species of insects and spiders which live in assoation with ants, although not all of these mimic ants. Much of the communication in ant and termite colonies is by smell and, in the same way that some animals have evolved visual mimicry for the purposes of entering ants' nests, it appears that a few have evolved chemical mimicry to overcome the ants' defences and are able to enter the nest without being recognized as an intruder and attacked. There are a number of species of rove beetle which, as well as resembling ants in shape and colour, release chemicals which cause the ants to mistake the beetle for one of their own species. The beetle can then enter the nest, where it may be fed by the ants, and is also able to prey on the eggs and larvae. A similar system of chemical mimicry appears to enable some ants to invade termite nests without being recognized as hostile.

Aphids are often tended by ants and it has been suggested that the rear end of an aphid mimics the head of an ant. The aphid moves its hind legs in a similar way to the waggling of an ant's antennae. An aphid touched in the rear by the antennae of an ant excretes honeydew which the ant consumes. Possibly the aphid's behaviour is an invitation to the ant to accept honeydew. Thus although aphids do not really look like ants it is possible that some of their behaviour is mimetic.

Assemblages of nasty beetles

In the tropics and the warmer parts of the temperate regions, particularly in the southwestern United States, large assemblages of brightly-coloured beetles are often found. The beetles sit motionless, or nearly so, near the top of low-growing vegetation, and when approached make no effort to escape. It is possible to find an assemblage where almost every piece of vegetation is covered with beetles, yet a few metres away in apparently identical surroundings no beetles are to be seen. Within an assemblage the beetles are all alike, but other assemblages may consist of beetles of a rather different coloration. The commonest colour combination is orange with bold black tips to the wing-cases, which immediately suggests warning coloration. Most of the beetles belong to the family Lycidae. They have broad, soft wing-cases and when handled feel limp, not hard like most beetles. Among these assemblages of lycids there are invariably other insects with a similar coloration. These are mimics and they comprise an extraordinary variety of species, with representatives from several quite different groups including moths, solitary wasps, sawflies, ichneumonids, flies, bugs and especially other beetles. In West Africa there are even one or two species of small, orange and black lycaenid butterflies that appear to mimic lycids. Using the same criterion as we used for *Vespula* wasps, it would appear that lycids are very nasty, for not only do they have many mimics but both models and mimics crowd together almost as if to make themselves as obvious as possible.

The nastiness of lycids results from what is called 'reflex bleeding'. When handled or when attacked they produce drops of their own blood from joints in the legs and from other parts of the body. The fluid is not squirted and there is no foam: it simply leaks out. Other insects, especially ants, and also birds and lizards are put off by the fluid and refuse to eat lycids. A beetle that has been attacked and rejected normally recovers and so the advantage of reflex bleeding is that the predator gets a taste but the prey is not killed. Some other groups of unpalatable beetles, including ladybirds, are also capable of reflex bleeding; in oil beetles of the family Meloidae the blood contains cantharidin which is deadly poisonous to humans.

Most of the mimics present in the assemblages of lycids are palatable insects but there are also some unpalatable Müllerian mimics, particularly moths. However it is doubtful if they are as nasty to eat as the lycids. In one assemblage studied in the United States two of the mimics, *Electroleptus ignitus* and *Electroleptus apicalis*, beetles of the family Cerambycidae, not only resembled lycids but also fed on them. They were observed to drink the blood, sometimes, but not always, killing the models. By drinking the blood of the lycids the mimics make themselves less palatable. There is no suggestion, as with the *Volucella* flies described earlier, of resemblance enabling access to a food source because the lycids do not defend themselves other than by reflex bleeding. Here we have a Batesian mimic temporarily converting itself into a Müllerian mimic by absorbing the distasteful ingredients of its model's blood.

Firefly femmes fatales

Despite their name fireflies are not flies but beetles belonging to the family Lampyridae. They are active at night, and males and females recognize each other by Morse-code type signals in the flash of their lights: each species has a distinctive flashing sequence. Females usually remain still on vegetation and answer the lights of passing males of

their own species. In the United States the females of the genus *Photuris* prey upon other insects among them the males of *Photinus*, another genus of fireflies. Female *Photuris* flash at passing *Photinus* males and attract them by mimicking the flashing of *Photinus* females. When a male alights beside a female *Photuris* it is seized and eaten. As is common among animals the male is lured to the female, but in this instance what had promised to be a mate turns out to be a vicious predator. Presumably male *Photuris* give appropriate (but at present unknown) signals to their own females which enable them to escape the fate of male *Photinus*. In this example the mimicry involves tricking the prey into mistaking a predator for a mate. This is one example of what is sometimes called 'aggressive mimicry'.

Mimicry in vertebrates

Examples of mimicry in birds and mammals are few and poorly understood; indeed most examples of what are claimed to be mimicry border on the anecdotal. Tropical drongos *Dicrurus* are black birds and their flesh is reputedly unpalatable; they are also aggressive and noisy. In many places where they occur there are black flycatchers and shrikes, which, it is claimed, are Batesian mimics of the drongos. In Africa two species of rufous flycatcher *Stizorhina* are extremely similar in coloration to two species of ant-thrush *Neocossyphus* which are also reputedly unpalatable, possibly because they eat ants and taste of formic acid derived from the ants.

Most salamanders are camouflaged above and, in the breeding season, brightly coloured below, the colours functioning in courtship displays. But the brightly coloured red and black North American newt *Notophthalmus viridescens*, is rejected when offered to chickens and is presumed to be unpalatable to its normal predators which are probably birds and snakes. It is

Both of these African birds are found in the same area. The fork-tailed drongo *Dicrurus adsimilis* (left) is believed to be the model for the mimetic southern black flycatcher *Melaenornis pammelaina*.

piece of their flesh torn away and eaten. The sabre-toothed blenny is thus a mimic of the sea swallow. The mimicry involves tricking the prey which mistake it for a cleaner fish, another example of aggressive mimicry.

The evolution of mimicry

The mimetic associations described in this chapter have in most cases taken thousands, maybe tens or hundreds of thousands of years to evolve and there is consequently little hope of obtaining much direct evidence for the evolution of mimicry. Presumably, the distasteful models and their warning coloration evolved before the mimics, and subsequently mimics gradually developed the appearance of noxious species. A sudden mutation from a non-mimetic to a mimetic coloration is unlikely to have occurred; more likely there were slight changes which gave small advantages to the mimic and as time passed these were improved upon by natural selection.

Various experiments have been performed to prove the effectiveness of mimicry and to show that predators learn and remember. In one experiment chickens avoided drinking from water that had been coloured green and from which they received a mild electric shock. Then when offered green water without the shock they would not drink it. When given various dilutions of green they drank more readily depending on how different the colour of the water was from the green water that gave the shock. In this experiment the green, electrified water is the model and the various shades of green the mimics. The chickens were unwilling to try mimetic water that closely resembled model water, but not so reluctant to try mimetic water that resembled the model water less accurately. This is analogous to what must occur with real models and mimics: the better the resemblance the more chance a mimic has of being avoided by a predator.

mimicked by a salamander, *Pseudotriton ruber*, which is rejected by chickens that have tried *Notophthalmus viridescens*, but readily eaten by chickens with no previous experience of the model.

Some species of small fish obtain food by cleaning off parasites and fungal growths attached to the bodies of larger species of fish. Cleaner fish are tolerated by predatory fish such as sharks that might otherwise make a meal of them, a good example of a mutually beneficial relationship. One cleaner fish, the sea swallow *Labroides dimidiatus*, occurs around coral reefs in the Pacific and Indian Oceans; it is territorial and fish of other species enter its territory to be cleaned. Sea swallows are boldly striped with blue, black and white, suggesting warning coloration, but they are not unpalatable: their distinctive coloration enables fish needing to be cleaned to recognize them. The sabre-toothed blenny *Aspidontus taeniatus* is almost identical in coloration to the sea swallow and is likewise territorial. Customer fish approach it but instead of getting themselves cleaned they are attacked and a

Other experiments have involved

Mimicry is not only visual but behavioural. The swimming movements of the carnivorous sabre-toothed blenny (right and enlarged) are similar to those of a cleaner fish, the sea swallow.

offering wild birds coloured pastry baits, flavoured with quinine to make them unpalatable, and then later offering pastry baits of the same or similar colours but without the quinine. The birds learn by experience, and the closer the coloration of model baits and mimetic baits the greater the reluctance of birds to try the mimics. Such experiments as these replicate what occurs in nature and add support to results from experiments in which predators are offered real models and mimics and learn to reject the mimics after they have tried the models. So, one requirement for the evolution of mimicry is well established: predators do learn to discriminate, they are easily confused and they remember previous unpleasant experiences. They are more suspicious of good mimics than poor ones, and thus are capable of exerting the selection needed for the evolution of a close resemblance between unpalatable and palatable species. It is apparently less easy to devise an experiment to show how quickly predators forget, but there is no doubt that in natural situations a predator's image of unpalatable or dangerous prey is reinforced by repeated encounters.

For mimicry to evolve it is of course essential that mimetic resemblance is inherited, and experiments with polymorphic butterflies show that their colour and patterns are indeed under genetic control. As Bates appreciated, and as we shall see in the next chapter, it is butterflies that provide the best insight into mimicry.

A confusing array of butterflies

Imagine a walk in a tropical forest. There are butterflies everywhere: large and gaudy, small and inconspicuous; they shimmer in the sunlight and disappear into deep shade. Some are restricted to the forest floor and now and again dense aggregations are encountered feeding from fallen fruit or imbibing from a patch of urine left by a passing monkey. Above ground at eye level there are different species, many boldly patterned with black and white, black and orange, or black and yellow. Here and there individuals are defending territories: they perch on a twig that gives them a good vantage point and intercept and see off others that enter their airspace. High in the canopy there are hordes of butterflies which can be seen only in silhouette. Many are small, probably lycaenids, but there are also large swallowtails. A dragonfly darts by, carrying a butterfly it has caught. A bird makes sorties from a perch to snap at flying butterflies; if it is successful it returns to the perch, eats the butterfly's body and allows the wings to flutter to the ground. A close look at the ground may reveal more wings that have been discarded by predators.

The array of colours and patterns is bewildering. Butterflies that look alike turn out on close inspection to belong to different families let alone different species, while some that look very different from each other may be found mating, showing that they are the same species. In Europe it is usually possible to identify a butterfly from its colour and pattern; in the tropics this is often not possible. Much of the difficulty arises because there are so many models and mimics and because of striking individual variation in coloration. There are also many species: three hundred or more in a West African forest, only about thirty in an equivalent area of English woodland. This is partly because a tropical forest is much richer than temperate woodland in species of trees and plants which are the food-plants of butterfly caterpillars: the more kinds of plants, the more butterflies, each of them feeding as caterpillars on a relatively narrow range of related plants.

A visitor's view of butterflies in a tropical forest contrasts markedly with that of an experienced butterfly-watcher. The butterfly-watcher can quite easily separate models from mimics by subtle differences in flight and he is aware of, rather than confused by, polymorphic colour forms. A predator's view is probably somewhere in between. Repeated unpleasant experiences remind it that certain butterflies are nasty and must be avoided and so good mimics are also avoided. The predator is cautious and suspicious and the result is that although many butterflies are caught and eaten, many escape as a result of the confusion brought about by mimetic deception.

In tropical forests, large clusters of butterflies are frequently found feeding on damp soil. Close examination of such a cluster reveals similar-looking individuals that are totally unrelated and different-looking individuals which belong to the same species. Much of the confusion arises because of mimetic resemblance.

Black and orange and black and white African butterflies

Throughout tropical African forests there are black and orange and black and white butterflies which are connected by similarity in coloration to form one of the most complex examples of mimicry known. The models are species of Danaidae and Acraeidae, some of which are Müllerian mimics of each other; the Batesian mimics are representatives from the families Nymphalidae, Papilionidae, Satyridae and Lycaenidae, as well as several species of day-flying moths. The two colour combinations are connected because in some species one sex is black and white and the other black and orange, and because several mimics are polymorphic with both black and orange and black and white colour forms. The species composition of the assemblage varies from place to place, but almost everywhere models are about five times as abundant as mimics.

In one well-studied patch of forest in Sierra Leone the complex involves no less than 18 species, with ten species acting as models, and eight species as mimics of one sort or another. The web of mimetic relationships between these species is immensely complicated. The models are four species of *Amauris* (Danaidae), five species of *Bematistes* (Acraeidae), and the females of *Acraea jodutta* (also Acraeidae). The mimics are two species of *Hypolimnas* and *Pseudacraea eurytus* (Nymphalidae), *Elymnias bammakoo* (Satyridae), *Mimacraea neurata* and *Pseudaletis leonis* (Lycaenidae), and the females of two species of *Papilio* (Papilionidae). Three of the species of *Amauris* are very similar in coloration with white-spotted, black forewings and black hindwings with a large yellowish-white patch nearest the body. They are mimicked by *Hypolimnas dinarcha* and one of the two colour forms of *Hypolimnas dubius*. The other colour form of *Hypolimnas dubius* is a mimic of the boldly patterned black and white *Amauris niavius*, which is also mimicked

by the females of *Papilio dardanus* and *Papilio cynorta*, one of the two colour forms of *Elymnias bammakoo*, both sexes of *Pseudaletis leonis* and a female-limited colour form of *Pseudacraea eurytus*. The females of two other models, *Bematistes epaea* and *Acraea jodutta*, closely resemble *Amauris niavius* and are therefore additional models for the Batesian mimics and, of course, this makes all three models Müllerian mimics. The females of two more models, *Bematistes macaria* and *Bematistes alcinoe*, are rather differently patterned with black and white and are mimicked by another female-limited colour form of *Pseudacraea eurytus*. The males of these two models are slightly different from each other but both are black and reddish-orange and are mimicked by a male-limited form and a form that occurs in both sexes of *Pseudacraea eurytus*. In *Bematistes vestalis* and *Bematistes umbra* the sexes are alike; they have dusky forewings and mainly orange hindwings and are mimicked by a form of *Pseudacraea eurytus* that occurs in both sexes. Finally we are left with the males of *Bematistes epaea* which are black with distinctive bright orange markings in the forewings and mainly orange hindwings. They are mimicked by identical-looking male and female forms of *Pseudacraea eurytus*, the second colour form of *Elymnias bammakoo*, and both sexes of *Mimacraea neaurata*.

Confusing? Yes, but this is the point. Here is an assemblage of butterflies connected together in a most intricate way, yet at the same time the colour patterns involved are variations on a simple theme. Note in particular that only the females of the two species of *Papilio* are mimics and that only the females of *Acraea jodutta* are models. Note also that *Pseudacraea eurytus* has male- and female-limited colour forms and colour forms that occur in both sexes; in this species the identical-looking male and female forms that mimic *Bematistes epaea* are produced by different genotypes, which is most unusual.

Pseudacraea eurytus is the most polymorphic of all mimetic butterflies. Samples of it and its models collected in forests throughout Africa show that very few individuals are non-mimetic. The abundance of each form is determined by the abundance of the models that happen to be present. Such an arrangement results from frequency-dependent selection in which mimetic forms that tend to become too common in relation to the abundance of their models are reduced in numbers by predators and so a balance is achieved which changes only when the models alter in abundance. *Pseudacraea eurytus* is common on the islands in Lake Victoria but here the models are scarce. Between 28 per cent and 56 per cent, depending on the island, are non-mimetic whereas on the mainland at Entebbe where the models are common only 4 per cent are non-mimetic.

Mimetic African swallowtails

In Sierra Leone nearly all the females of the swallowtail, *Papilio dardanus*, are black and white mimics of the monarch, *Amauris niavius*. Further east, especially in the Congo Basin and in Uganda, several different colour forms occur together, each a mimic of a species of *Amauris* or *Bematistes*. But everywhere in Africa where the butterfly occurs, the males are pale yellow with black markings and are not mimetic.

The restriction of mimetic colour forms to females occurs in other species of tropical swallowtails and in some mimetic nymphalids. In the North American tiger swallowtail, described in Chapter 9, the black female mimetic form seems to be inherited via the Y chromosome. But this is unusual, and in other species where mimetic polymorphism is restricted to females the mimetic pattern is determined by genes present in both sexes but which only express themselves in the females. This is called sex-limited or sex-controlled inheritance. The phenotypes of the female offspring are therefore dependent on the genotype of both parents, not just on the genotype of the female as in the tiger swallowtail. Why should mimetic polymorphism be restricted to females in so many species? There are several possibilities, but the most likely is that coloration also plays a part in courtship behaviour and so the males retain the distinctive coloration of the species which can be recognized by the females. Another possibility is that females have to locate and lay eggs on the right food-plant for the caterpillars to feed on and so are much more vulnerable to predators and need better protection.

Male and female *Papilio dardanus* can be mated in captivity and in this way the inheritance of mimetic forms has been worked out. The forms are controlled by multiple alleles which produce dominant and recessive phenotypes in predictable ratios. Matings between butterflies from the same area result in offspring that are good mimics of the local models. But matings between butterflies from widely separated places produce a variety of intermediate-looking phenotypes in the offspring, many of which are poor mimics and, moreover, do not resemble any form of *Papilio dardanus* found in nature. Natural selection by predators would soon eliminate such forms if they were to occur in the wild. Such selection has led, in each locality, to the evolution of modifier genes, which control the expression of genes affecting colour and pattern so that either one colour form or another is produced, and not an intermediate. When butterflies from different localities are crossed the modifier genes cannot act effectively and intermediates are produced.

Papilio dardanus is not mimetic everywhere: on Madagascar, where suitable models are few, the females are non-mimetic and male-like. And, as in *Pseudacraea eurytus*, there are places where the mimics outnumber the models and the mimicry is much less perfect.

Mimicry in West African butterflies. The models are all members of the families Danaidae and Acraeidae. The sex is given only in cases where there is a difference in colour and pattern.

Amauris egialea Danaidae
Amauris hecate Danaidae
Amauris tartarea Danaidae
Hypolimnas dubius a colour form
Nymphalidae
Hypolimnas dinarcha Nymphalidae

Bematistes macaria Acraeidae
Bematistes alcinoe female Acraeidae
Pseudacraea eurytus a female colour form
Nymphalidae
Bematistes macaria male Acraeidae
Pseudacraea eurytus a male colour form
Nymphalidae

Bematistes vestalis Acraeidae
Pseudacraea eurytus a colour form
Nymphalidae
Bematistes umbra Acraeidae
Bematistes alcinoe male Acraeidae
Pseudacraea eurytus a colour form
Nymphalidae

Mimicry in West African butterflies. The models are all members of the families Danaidae and Acraeidae. The sex is given only in cases where there is a difference in colour and pattern.

Bematistes epaea male Acraeidae
Pseudacraea eurytus a male colour form Nymphalidae
Pseudacraea eurytus a female colour form Nymphalidae
Elymnias bammakoo a colour form Satyridae
Mimacraea neurata Lycaenidae

Bematistes epaea female Acraeidae
Acraea jodutta female Acraeidae
Pseudacraea eurytus a female colour form Nymphalidae
Elymnias bammakoo a colour form Satyridae
Papilio cynorta female Papilionidae

Amauris niavius Danaidae
Papilio dardanus a female colour form Papilionidae
Hypolimnas dubius a colour form Nymphalidae
Pitthea famula Geometridae
Pseudaletis leonis Lycaenidae
(*Pitthea famula* is a day-flying moth which appears to be part of the assemblage.)

The survival value of dying

One of the curiosities about many warningly coloured butterflies, including all species of monarchs, *Amauris*, *Danaus*, and related genera, is that there is no way a predator can learn that they are unpalatable unless it kills and tastes them. This is how predators learn to avoid models and their mimics when they encounter the same coloration again. But in this learning process the individual is sacrificed, and on first consideration it looks as if the prey's death can only be beneficial to other similarly coloured individuals. A warningly coloured butterfly that is killed is therefore in much the same category as a hero who sacrifices himself in the interests of others. Are butterflies heroic? No, and for the following reasons.

Many of the larger tropical species are long-lived and tend to remain in the same general area for weeks or months on end. We know this because individuals have been caught, marked on the wing with a spot of quick-drying ink, released, and subsequently recognized. This means that they are present during the period their offspring are developing; indeed in some species the caterpillar and pupa stages last only a short time – a matter of a few weeks – so that parents and offspring, possibly even grandchildren, may fly together. The offspring may therefore benefit from the death of a parent or grandparent because in this way the predators find out that they are unpalatable. The offspring carry some of the genes of the parents (or grandparents) and so apparently altruistic behaviour is nothing more than an individual protecting its own kind. This type of selection – where an individual dies in the interests of its relatives – could be called 'gene protection' but is usually referred to as 'kin selection'.

Some unpalatable butterflies, the Acraeidae, produce a foamy secretion from thoracic glands providing the predator with a chance of tasting, before killing, an individual it has caught. Like the lycid beetles and other insects that deter predators by producing an unpleasant smell or taste, they often escape with only minor injury.

All of this means that some mimetic butterflies rely for protection on a model's ability to produce a secretion while others rely on the model dying in the interests of its offspring.

Passion flower butterflies

There are about forty species of *Heliconius* butterflies in the forests of South and Central America. All have long, narrow, brightly-coloured wings, and all are unpalatable to predators. The caterpillars feed on the leaves of plants of the family Passifloraceae, often called passion flowers, and since none of the species has a generally accepted English name they are collectively called passion flower butterflies. It is these butterflies that first put Henry Walter Bates on to the theory of mimicry.

Passion flower butterflies are long-lived, and individuals tend to stay in the same small area of forest, often roosting communally at night. In some places there are as many as ten breeding generations in a year which means that parents, children, and grand-children regularly fly together. Many species of butterflies, and some day-flying moths, are Batesian mimics of passion flower butterflies. Two species, *Heliconius melpomene* and *Heliconius erato*, exhibit truly remarkable geographical variation in coloration. Nearly everywhere the two occur together, and in any one locality they look almost identical; indeed the best way of separating them is by the smell produced from their scent glands, but there are other small differences, including the markings of the caterpillars. The geographical variation involves different arrangements of black, red, yellow and blue patterns on the wings to an extent that butterflies of the same species from widely separated localities look like totally different species. The geographical variation is thus much greater than the difference between the two species. How can we account for this?

The resemblance between the two in any one place is Müllerian mimicry, but the geographical variation must be explained in another way. To do this we must digress and say a few words about the origin of geographical variation and the formation of races and species. In sexually reproducing animals, such as butterflies, a species is defined as a group of individuals that breed among themselves but do not naturally interbreed with other similar individuals; that is to say, a species is defined in terms of reproductive barriers between it and other species: *Heliconius melpomene* and *Heliconius erato* are separate species because they do not interbreed. New species can evolve when a population of one species becomes split into two or more parts that are geographically isolated from one another. Once isolated the populations can become adapted through natural selection to the environments where they are now restricted. Since these environments are geographically separated they will tend to differ in certain respects, in climate perhaps, and given time the populations become adapted to these conditions. If later on the populations meet because the geographical barriers separating them have ceased to exist, several possibilities arise. First, they may, on renewing contact, interbreed and in doing so become a single population again. If this happens nothing is left for the contemporary observer except a single interbreeding population and there is no evidence of what has happened in the past. Secondly, they may be unable to interbreed because they have evolved genetic differences that prevent them from doing so. In such situations the contemporary observer may see two or more similar species occupying much the same habitat and range. Thirdly, something in between these two possibilities occurs. The populations meet and where they do so they hybridize freely so that there are zones where several forms are found together and the species appears to be polymorphic. The species retains its distinctive racial characteristics in areas where these have evolved and loses them, or tends to lose them, where there are hybridization zones. This last possibility is probably what has happened to both *Heliconius melpomene* and *Heliconius erato*. We cannot of course be certain, as such a process takes thousands of breeding generations, and all we have available for study is the present situation. There are, in both species, zones of hybridization between races which certainly supports the hypothesis. Geographical differentiation in the numerous races of the two species probably occurred during the Ice Age when the forests of South and Central America were much more fragmented than they are nowadays. In the warm periods between the successive glaciations, known as interglacial periods, the climate of the American tropics was undoubtedly drier than it is now and this must have meant less extensive forest and, more important, many small and isolated blocks of forest. Given this situation, *Heliconius melpomene* and *Heliconius erato* evolved and differentiated but as the climate got warmer their geographical ranges expanded again and previously isolated populations met.

Hence natural selection has had two effects on the coloration of these two species of *Heliconius*. It has produced similar mimetic coloration in response to selection by predators, and has led to the evolution of races. We do not know why the races are so strikingly different from each other but, whatever the cause, it has produced the same effect in both species. Geographical races of the sort found in these two butterflies are an example of incipient speciation: had more time elapsed while they were isolated they might have evolved into many distinct species.

The passion flower butterflies also have some interesting co-evolutionary relationships with the food-plants of their caterpillars. The passion flowers have evolved many defences that protect them from caterpillars and from egg-laying butterflies. One is that some of the species of *Passiflora* produce swollen, yellow projections on the leaves and tendrils which look exactly like *Heliconius* eggs. Female butterflies avoid laying where eggs are already present and so the plants of these species receive few eggs and are less damaged by caterpillars.

Small yellow projections on the leaves and tendrils of a passion flower plant resemble the eggs of *Heliconius* butterflies and deter females from laying. The lowest yellow projection is a real butterfly egg.

Mimetic butterflies of the African savanna

The African monarch *Danaus chrysippus* is found in grassy savanna and, nowadays, in cultivated places in the forest region. The caterpillars feed on milkweeds and, it will be recalled, absorb cardenolides which make the butterflies unpalatable. The considerable variation in cardenolide content in the butterflies depends partly on the species of milkweed eaten and partly on individual differences in the ability of the caterpillars to absorb and store the compounds. The proportion of unpalatable individuals varies from place to place.

The African monarch is the key species in a mimetic assemblage of great complexity. Unlike most models it is polymorphic in coloration in many parts of its range. There are four colour forms, each of which has received a scientific

Opposite Mimetic butterflies of the African savannah.
1 *Acraea encedon* a colour form
2 *Acraea encedon* another colour form
3 *Danaus chrysippus* form *dorippus*
4 *Hypolimnas misippus* female, a colour form
5 *Danaus chrysippus* form *chrysippus*
6 *Hypolimnas misippus* female, a colour form
7 *Hypolimnas misippus* male, non-mimetic
The females of *Hypolimnas misippus* are Batesian mimics of *Danaus chrysippus* and possibly *Acraea encedon*. *Danaus chrysippus* and *Acraea encedon* are Müllerian mimics of each other.

name, and for convenience we shall use these names. Form *chrysippus* is orange with broad, white-spotted black tips to the forewings; *alcippus* is the same but the hindwings are mainly white; *dorippus* is entirely orange and lacks the wing-tip pattern; and *albinus* is like *dorippus* but with a mainly white hindwing. The African monarch is not polymorphic in all parts of its range. In much of West Africa only *alcippus* occurs, and in much of north and southwest Africa and on Madagascar there is only *chrysippus*. In the northeast horn of Africa *dorippus* is probably the only form. In Central and East Africa all four forms are found together but their frequency varies, *dorippus* becoming increasingly common towards the east coast and *alcippus* from Uganda westwards.

The nymphalid butterfly, *Hypolimnas misippus*, is a Batesian mimic of the African monarch and is also polymorphic, having four forms which correspond to those of the monarch. In this species mimicry is confined to the female, the males being black with a large white spot on each of the wings. At Dar es Salaam the relative frequency of the four forms of mimic rarely matches that of the model and here, as elsewhere in Africa, intermediates are much more common in the mimic than in the model. In West Africa all four forms and many intermediates of the mimic are found, although only the *alcippus* form of the monarch occurs. Paradoxically, the form which mimics *alcippus* is rare.

In many other parts of Africa and on Madagascar *Hypolimnas misippus* is more polymorphic than the monarch and although the occurrence of the monarch's four colour forms show broad geographical trends, those of the mimic do not. Nearly always the mimic is less abundant than the model, but in places with markedly wet and dry seasons the model is more abundant in the dry season and the mimic more abundant in the wet season. This is not a very impressive association on the part of a butterfly

believed to be a Batesian mimic: what are we to make of it?

Before attempting an explanation it is necessary to mention that other species are involved in the assemblage.

Pseudacraea poggei is a rather rare nymphalid found in wooded savanna in southern and eastern Africa. It looks like and flies in the same buoyant way as the African monarch and mimics both the *chrysippus* and the *dorippus* forms. Like its extremely polymorphic forest relative, *Pseudacraea eurytus*, there is little doubt that it is a Batesian mimic. One of the female forms of the swallowtail, *Papilio dardanus*, is a rather poor mimic of the *chrysippus* form of the model. *Papilio dardanus* is a forest butterfly and although it occurs in cultivated places and gardens where the African monarch is also found, the habitats of mimic and model largely separate them from each other. There are, in addition, several species of forest nymphalids, including the females of *Euriphene atossa*, which have a *chrysippus*-like appearance, and several species of day-flying arctiid moths have the same general coloration. Little is known about this part of the assemblage which seems to associate forest butterflies with savanna butterflies, but the chances are that the moths are the models and that in reality there are two assemblages.

Then there is *Acraea encedon*, a polymorphic Müllerian mimic of the African monarch, found throughout Africa south of the Sahara where it tends to form discrete, isolated populations. Each population is characterized by the number and relative frequency of a host of colour forms, some of which are not mimetic, while there are also populations which are neither mimetic nor polymorphic. In Uganda four forms of *Acraea encedon* are mimics of the four forms of the African monarch but, depending on the population, there are always non-mimetic forms varying in frequency between 3 per cent and 38 per cent. In Sierra Leone most of the *Acraea encedon* do not mimic the African monarch, but in one population (which may in fact be of a different species) all are excellent mimics of the *alcippus* form. It is known that *Acraea encedon* is unpalatable but it is not known if palatability varies between individuals and from place to place.

We can now return to what all this means. The mimetic assemblage surrounding the African monarch has been more carefully studied than any other and as a consequence new and intriguing problems in ecology and genetics have come to light. The chances are that the butterfly is more unpalatable in East Africa, where it is also more polymorphic, than in West Africa where it is not polymorphic. It also has more mimics, Batesian and Müllerian, in the east than in the west; indeed in West Africa mimicry is poorly developed and in some places non-existent.

A model with many mimics is in danger because predators are likely to find the mimics too often and consequently attack the model. One way for the model to escape is to evolve contrasting polymorphic forms, as the African monarch seems to have done in East Africa. This 'off-loads' some of the mimics which are no longer a good match, but the strategy is countered by the evolution of matching polymorphic forms in the mimics. This is what seems to have happened in *Hypolimnas misippus* and *Pseudacraea poggei*, and possibly in the Müllerian mimic *Acraea encedon*. Hence one hypothesis is that polymorphism in models is an adaptation that to some extent lessens their dangerous burden of mimics. *Acraea encedon* is possibly more unpalatable than the African monarch and at least in some places the species we have identified as the key model in the assemblage is a Batesian as well as a Müllerian mimic.

For most butterflies it is assumed, probably incorrectly, that mates are randomly chosen and that there are no preferences. Polymorphic species like the African monarch provide an opportunity to examine this assumption. At Dar es Salaam there is a marked tendency for unlike *chrysippus* and *dorippus* forms to mate with each other. Such a preference may be facilitated by the different appearance of the forms; in other words the polymorphism may have been evolved, at least in part, as a means of enabling mating between unlike genotypes. This discovery led to further investigation and it turns out that the mating advantages of the two forms vary with the season and with their frequency in the population; for example, at certain times of year *chrysippus* males have better mating prospects than *dorippus* males which later leads to an increase in the frequency of the *chrysippus* form. These results indicate that ploymorphism in the African monarch may not necessarily have been evolved solely as a means of protection from predators. The forms may have several quite different functions and this may be one reason why there is not always a good correlation between the occurrence and frequency of model and mimetic forms in butterflies.

There is one final point. In Africa the grand design of forest and savanna has been drastically disrupted by the expanding human population: forests have been and are being destroyed and everywhere cultivation has spread and altered the natural landscape. Most of the mimetic butterflies discussed in this section have exploited the changed circumstances. They have spread into farmland and into gardens and now occur throughout the forest region wherever trees have been felled and the land cleared for cultivation. They have even switched to non-native food-plants. In the case of the African monarch these include milkweeds from Central America grown as garden ornamentals; no doubt this has changed the frequency of palatable and unpalatable individuals. All of this has happened in the last two hundred years or so and we are now looking at a gigantic mix-up of models and mimics which does not always make sense.

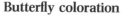

Another mimetic relationship:
1 *Papilio glaucus* mimetic female
2 *Papilio glaucus* male, non-mimetic
3 *Limenitis arthemis* mimic
4 *Limenitis arthemis* non-mimetic, outside the
   range of the model
5 *Battus philenor* model
The mimetic relationships of these butterflies are
described, in Chapter 9, in mimicry in North
American butterflies.

Butterfly coloration

Butterflies are among the most colourful of all animals. We know that colour patterns are used as a means of individuals recognizing others of their species and that many of the differences between species have been evolved to minimize 'mistakes' in courtship and display. Coloration is also important in temperature regulation – there are experiments that show this – and in sexual recognition, camouflage, distracting and intimidating predators and, as we have seen in this chapter, in warning coloration and mimicry. If we are not careful we can forget that a butterfly's wings are also used for flying. They must have the right aerodynamic properties for flapping, gliding, twisting and turning, and they must be resistant to stress and wear. Black wing tips, for example, occur in many species, and may well strengthen the wing and reduce wear. Indeed some of the striking warning and mimetic colour patterns may have first evolved for quite different reasons and the likelihood is that some of the types of coloration described in this chapter have additional but as yet unknown functions.

11 The tangled web of deception

The anti-predator strategies discussed in this book could not have evolved and would not have persisted in the absence of predators. If nothing preyed upon camouflaged grasshoppers the camouflage would disappear, and warning coloration exists because individuals possessing it are still attacked and eaten, although less often than if they did not have warning coloration. The strategies of predators and prey have evolved together, reminding us of the relationship between flowers and their pollinators: one is not possible without the other. We have, for convenience, considered each strategy separately, starting with camouflage and ending with mimicry, taking examples from throughout the animal kingdom, and from time to time digressing to include similar strategies in plants. This approach has certain deficiencies. Even a single individual may use several different deceptive and defensive tricks when confronted by a predator, just as a predator may use several different ways of finding and overcoming its prey.

Take, as an example, the fully-grown caterpillar of the European puss moth *Cerura vinula* which feeds at night on the leaves of willow and poplar and rests motionless by day on a leaf or stem. It is pale green with a white-edged, diamond-shaped saddle of purplish-brown on the back. Despite its large size and robust appearance the caterpillar is well camouflaged: the green resembles the green of leaves and the saddle breaks up its shape. But it must keep still. If it does not, and is discovered, a fold of skin is inflated around the head; at the same time the head is withdrawn a little and a formidable-looking scarlet 'face' is produced on which there are two black spots resembling eyes. As this is happening the caterpillar raises both its front and rear ends and becomes boat-shaped. If the predator is not put off by this behaviour the two rear tails protrude bright red, whip-like threads which are lashed about in a menacing way. The head is rocked from side to side and food is regurgitated from the mouth to produce a slightly acrid smell. Even if the predator proceeds all is not lost. The shield-like area behind the head is slippery and quite hard and since this is the region where a predator such as a bird might make its first strike, the blow is softened or perhaps deflected. The caterpillar also tightens its grip on the leaf or twig and, because of its large size, is difficult to dislodge. The puss moth caterpillar therefore uses camouflage and disruptive coloration for primary defence and various forms of intimidation as well as sheer strength for secondary defence. It never gives up but sits out an attack by responding with a barrage of defences until the predator goes away or, of course, until it is eaten. A small bird that finds and eventually overcomes one is worthy of respect.

The stick insects, or walking-sticks, and leaf insects (Phasmida) are a fine example of a group of insects within which occurs a variety of deceptive and defensive strategies. There are about two thousand species and all feed on green leaves and, to a lesser extent, on stems and flowers. Most species are tropical. They are beautifully camouflaged and resemble the vegetation among which they occur. Some species have green and brown colour forms and, like grasshoppers, can select the appropriate background on which to rest.

When provoked a phasmid stops moving and 'freezes'. Its resemblance to the vegetation is so exact that if you take your eyes off it for a moment it may be diffcult to find again. In most species immature individuals look like smaller versions of the adults. In a few, including the Australian Macleay's spectre *Extatosoma tiaratum*, they are ant mimics when very young; not only do they look like ants but run around rapidly in much the same way and are presumably avoided by the many predators that do not like ants. As they get bigger they moult and lose their ant-like appearance and become like diminutive adults. The eggs of phasmids seem to resemble small seeds: a good example of looking like something different. These are all ways of primary defence from predators.

If discovered and attacked some species, including Macleay's spectre, grab at the predator with their front legs and at the same time curl the abdomen upwards to give a scorpion-like appearance. This is threatening and intimidating and presumably works well as many phasmids are found in the same places as dangerous scorpions. Some species suddenly open their wings when attacked and expose bright patches of colour, behaviour reminiscent of underwing moths. The Javanese lichen stick *Orxines macklottii* looks like lichen and is almost impossible to see when on

the right background; if disturbed it opens its wings with a rustle, showing off patches of orange bordered with black and white, and produces an unpleasant smell. The Florida stick *Anisomorpha buprestoides* sprays a defensive fluid backwards from special glands in the thorax. The spray is accurately aimed and usually hits the attacker in the face, a strategy similar to that of the bombardier beetle. All phasmids are capable of shedding legs that have been seized by a predator, reminding us of lizards which shed their tails. Unlike most insects, immature phasmids can regenerate lost legs; adults are not able to do this but provided they are left with some legs they can still function quite well. These are all examples of secondary defence.

In this single group of insects there are many of the deceptive and defensive strategies found in the animal kingdom. About a dozen species of phasmids are readily available from dealers in live insects and they are easy to rear in a temperate climate if they are kept indoors. They are well worth keeping and observing.

There is plenty of information about what prey species predators take; for birds in particular there are detailed inventories of the food items eaten in the wild, and there is a considerable body of information on the food of mammals, reptiles and fish, but rather less for predatory invertebrates like spiders, mantids and dragonflies. But we know little about the success and failure rates of predators and very little indeed about why certain individuals are killed and others missed or overlooked. In the 1930s and 1940s painstaking recordings were made of the frequency of successful and unsuccessful hunts by several species of birds of prey. One observer witnessed 213 hunts by sparrowhawks, of which only 23 resulted in a kill: a very low success rate. What is more interesting is that the successes resulted because the birds attacked were either already injured or

Half-grown stick insects, *Extatosoma tiaratum,* which look like leaves. When they are younger they look like ants.

were behaving in an abnormal way. It was suggested that in the sparrowhawk and in other birds of prey the success rate is not only low but that a sizeable proportion of the victims have made mistakes or are in the wrong surroundings.

Herons are essentially fish-eaters but fish are not always easy to find and catch. Rather than go hungry they will turn to other prey. They will catch and eat newly emerged dragonflies while these are still limp and incapable of escape by flying away. Dragonflies are scarcely satisfying food for large birds like herons but what else can an individual do if it cannot catch fish? We think of lions as fierce predators of antelopes but their success rate is low and they are frequently forced to hunt birds and even frogs. The boldest of predators can

Left Fully grown puss moth caterpillar in normal resting position.

Right Puss moth caterpillar defending itself by producing a face and lashing its whip-like tail threads. These are but two of the defensive strategies the caterpillar is able to employ.

No matter how well stick insects resemble sticks, large numbers will still get eaten by predators.

be humbled at a time of food scarcity, which may be the rule rather than the exception.

All the same, we can envisage predators as opportunists which are quick to respond to mistakes made by prey, but compared to an experienced naturalist they are not especially competent. A naturalist can recognize signs and then search for his quarry. Fresh holes in leaves and the presence of droppings indicate that caterpillars are in the vicinity; sticky honeydew means aphids; the footprints of mammals and birds tell which species are about and how recently they have passed by; and the details of wing venation quickly separate model from mimic. No predator possesses these abilities. All rely on trial and error searching, a certain amount of learning from experience,

and all face the prospect of repeated failure. They are easily fooled, tricked and intimidated, and are quick to give up and try again elsewhere.

The number and complexity of niches and the organisms that operate within them is astounding. Each species has available a whole battery of survival strategies. The web of adaptation is vast and we are only beginning to understand how natural selection has moulded plants and animals to fit into their surroundings. Deception is an essential way of life and the very complexity of the ways in which it is achieved must lead us to marvel at the intricacies of the living world.

Glossary

abdomen in adult insects and spiders, the posterior section of the body which lacks appendages. See also **thorax**

aggressive mimicry resemblance of a predator (or parasite) to a harmless animal to facilitate approach to the prey (or host)

algae simple, predominantly aquatic plants; many are single celled and visible only under the microscope, although their effect in turning water green may be visible to the naked eye

allele an alternative form of a gene at a given locus (*qv*)

amphibians a class of vertebrates including the frogs, toads, newts and salamanders

antennae in insects, paired appendages of the head which have a sensory function, being sensitive to touch, smell, and in some species sound

automimicry a form of mimicry where mimics and their models are members of the same species

Batesian mimicry the resemblance of a palatable mimic to an unpalatable model

bivalves a class of molluscs characterized by a hinged double shell, *eg* clams, oysters

bugs members of the order Hemiptera, a group of insects with piercing mouthparts, *eg* shieldbugs, capsids, pondskaters

carapace a hard shield covering part of the body

caterpillar the larva (*qv*) of moths and butterflies; the larvae of sawflies are also sometimes known as caterpillars

cephalopods a class of molluscs characterized by sucker-bearing arms, *eg* octopuses, squids, cuttlefish

chromosomes thread- or rod-like bodies carrying the genetic material and located in the cells of organisms

community   the assemblage of species of plants and animals that occurs in a particular place

competition   the interaction of individuals or species resulting from use of essential resources that are in short supply

countershading   coloration which obscures the appearance of shadow on the body by having the underside paler than the upperside

crustaceans   a class of invertebrates which includes the crabs and shrimps

disruptive coloration coloration which disguises the body of an animal by breaking it up into visually distinct parts

diurnal   active by day

dominance   the mechanism by which one allele (*qv*) suppresses the expression of another allele

dorsal   of, or pertaining to, the back

enzyme   a substance that initiates or facilitates a chemical reaction between other substances

evolution   a cumulative, inheritable change in a population

family   in biological classification, a group of related genera

fitness   the relative ability of an organism to transmit its genes to the next generation

flash coloration   the bright coloration of parts of an animal which, though normally hidden, can be exposed suddenly to deter a predator

food web   an abstract connection between a group of species in a community (*qv*) describing which ones feed on which

gastropods   a class of molluscs characterized by a single, usually coiled, shell and a muscular foot, *eg* periwinkles, whelks, snails and slugs

gene   the smallest, indivisible unit of heredity

genetics   the study of heredity

genotype   the genetic constitution of an organism

genus (pl. **genera**) in biological classification, a group of related or similar species

hybridization   interbreeding or crossing

incomplete dominance the failure of one allele (*qv*) to suppress completely the action of another

invertebrate   animal lacking a backbone, *eg* worms, insects, crustaceans, molluscs

larva   the young stage of an animal if its characteristics differ considerably from those of the adult; in sedentary animals the larva is free-living

locus   a given site on a particular chromosome

mammals   a class of vertebrates that suckle their young with milk and generally have fur

metamorphosis   in insects and amphibians, the change from the larval to the adult form

midrib   the large central vein in a leaf

mimic   an organism which closely resembles another, usually unrelated, organism; the organism which it resembles is known as the model and is unpalatable. See also **aggressive mimicry**

model   an organism which is imitated by a mimic (*qv*)

modifier   a gene which modifies the effects of another gene

molluscs   a group of invertebrates including clams, octopuses and snails; they are predominantly shelled and, with the exception of land snails and slugs, largely aquatic. See also **bivalves, cephalopods** and **gastropods**

Müllerian mimicry resemblance between species when both have characteristics which are unpleasant to predators

mutation   a sudden change in the genetic material

natural selection the non-random elimination of individuals (and therefore of genes) from a population

niche   the place and role of an organism in its environment

operator   the organism that mimicry acts to deceive

parthenogenesis   reproduction by a female without fertilization by a male

phenotype   the physical characteristics of an organism

plankton   microscopic animals and plants that drift suspended in rivers, lakes and oceans

polygenic inheritance the determination of characteristics by the cumulative effects of a number of genes

polymorphism   the sustained appearance of two or more, genetically determined, distinct forms within a population

population   a group of organisms of the same species living and breeding together

proboscis   in invertebrates, a tubular mouthpart

pupa   in insect metamorphosis (*qv*), the reorganization stage that gives rise to the adult; the pupa is generally immobile

recessive   a genetic characteristic that is masked by its dominant counterpart. See also **dominance**

reptiles   a class of vertebrates including the lizards, snakes, tortoises and crocodiles

reverse countershading coloration in which the back is lighter in colour; found in animals which habitually rest upside-down. See also **countershading**

search-image   a predator's pictorial memory of the appearance of prey

speciation   the evolution of species

species   a group of individuals which show similar features and which can interbreed to produce viable offspring

territory   an area defended by an individual against other individuals of the same species

thorax   in insects and spiders, the region between the head and the abdomen (*qv*), bearing the legs and, in winged insects, the wings

ventral   the part of the animal normally facing the ground

vertebrate   animal possessing a backbone, *ie* fish, amphibians, reptiles, birds and mammals

Further reading

The classic work on deceptive colours and patterns is *Adaptive Coloration in Animals* by H. B. Cott (Methuen, 1940). This well illustrated book is long out of print but copies occasionally may be found in libraries. Much of the experimental research on camouflage, warning coloration, and mimicry has been done in the last thirty years, and so does not feature in Cott's book.

In writing my book I have relied heavily on articles published in scientific journals. Many of these articles are summarized in *Defence in Animals* by M. Edmunds (Longman, 1974) and *Mimicry in Plants and Animals* by W. Wickler (Weidenfeld and Nicolson, 1968), both illustrated in colour and black and white. There are excellent illustrations of predators and prey in *Colour for Survival* by P. Ward (Orbis, 1979) and in *The Hunters* by P. Whitfield (Hamlyn, 1978). The hunting strategies of predators are discussed in *Behavioural Ecology: an Evolutionary Approach* edited by J. R. Krebs and N. B. Davies (Blackwell, 1978) and *Animal Behavior: an Evolutionary Approach* by J. Alcock (Sinauer, 1975).

To understand natural selection it is necessary to have some background in genetics. In *Natural Selection and Heredity* by P. M. Sheppard (Hutchinson, 1967) the relationship between genetics and natural selection is explained with the aid of examples from melanic moths, polymorphic snails, and mimetic butter-flies. Also recommended are *Adaptation and Natural Selection* by G. C. Williams (Princeton University Press, 1974) and *The Selfish Gene* by R. Dawkins (Oxford University Press, 1976).

The phenomenon of industrial melanism in moths is described in detail in *The Evolution of Melanism* by H. B. D. Kettlewell (Oxford University Press, 1973). The various protective strategies of North American underwing moths are beautifully described in *Legion of Night* by T. D. Sargent (University of Massachusetts Press, 1976). Protective strategies in North European butterflies and moths are described in two books by E. B. Ford, *Butterflies* (Collins, 1977) and *Moths* (Collins, 1955). In *The World of Spiders* (Collins, 1971) W. S. Bristowe writes delightfully about experiments with crab spiders and the flowers which they match in colour, while various forms of deception found in marine molluscs are described and illustrated in *Living Marine Molluscs* by C. M. Yonge and T. E. Thompson (Collins, 1976). Mimicry in African butterflies, including the polymorphic species discussed in this book, is analyzed in *Tropical Butterflies* by D. F. Owen (Oxford University Press, 1971). Finally, *Stick and Leaf Insects* by J. T. Clark (Shurlock, 1978) tells how to rear these interesting insects in captivity and what protective strategies to look for in them.

Acknowledgements

11 Michael Tweedie; 13 Dr T. E. Thompson;
14 Jane Burton/Bruce Coleman Ltd; 15 Jane
Burton/Bruce Coleman Ltd; 20 Heather Angel;
21 Michael Tweedie; 24 Günter Ziesler;
25 Eric Hosking; 26 M. D. England/Ardea
London; 28-9 P. H. Ward/Natural Science
Photos; 30 Nicholas Hall; 32 Anthony
Maynard; 35 Michael Tweedie; 36 Michael
Tweedie; 37 top: Denis Owen; bottom: Sir
Cyril Clarke; 38 Michael Tweedie; 39 top:
Oxford Scientific Films; bottom: M. P. L.
Fogden; 40 top: Harold Schultz/Bruce
Coleman Ltd; bottom: Michael Tweedie;
41 Michael Tweedie; 42 Michael Tweedie;
43 H. J. Vermes/Professor T. D. Sargent, by
kind permission of the University of
Massachusetts Press; 44 Heather Angel;
45 Michael Tweedie; 47 Oxford Scientific Films;
48 Heather Angel; 49 Jennifer Owen;
50 Heather Angel; 54 M. P. L. Fogden;
55 Nicholas Hall; 56-7 Eric Hosking; 58 Eric
Hosking; 59 Eric Hosking; 60 Heather Angel;
61 Denis Owen; 62 M. P. L. Fogden; 65 M. P.
L. Fogden; 66 top: M. P. L. Fogden; bottom:
Professor Lincoln P. Bower; 67 top: M. P. L.
Fogden; bottom: B. A. Bowman/Natural
Science Photos; 68 Günter Ziesler; 70 top:
Heather Angel; bottom: Laurence Gould/
Oxford Scientific Films; 71 Peter Castell/Aquila;
74 Heather Angel; 75 top: Heather Angel;
bottom: Nicholas Hall 76 Nicholas Hall;
79 top: Michael Tweedie; bottom: Nicholas
Hall; 80 Michael Tweedie; 82-3 Eric Hosking;
84 Heather Angel; 86-7 Oxford Scientific Films;
89 H. J. Vermes/Professor T. D. Sargent, by
kind permission of the University of
Massachusetts Press; 90 Heather Angel;
91 Heather Angel; 93 Dr Ivan Sazima;
94 P. H. Ward/Natural Science Photos;
95 P. H. Ward/Natural Science Photos;
96-7 Michael Tweedie; 98 Heather Angel;
99 Richard Mills/Aquila; 100 Anthony
Maynard after Professor Thomas Eisner;

103 Jane Burton/Bruce Coleman Ltd;
106 Heather Angel; 107 David Duthie;
109 Heather Angel; 110-1 Norman Myers/
Bruce Coleman Ltd; 112 Michael Tweedie;
119 Hilary Burn; 120 Hilary Burn; 121 Hilary
Burn; 124 Hilary Burn; 125 Anthony
Maynard; 127 Heather Angel; 130 Denis
Owen; 131 Denis Owen; 134 Dr L. E. Gilbert;
135 Andrew Atkins; 138 Andrew Atkins;
141 Jane Burton/Bruce Coleman Ltd;
142 Heather Angel; 143 Heather Angel;
144 Heather Angel.

Index